GW00382278

ONE HORSEPOWER

Logging in the Lake District 1977-1984

and other horse tales

by Bill Lloyd

A long walk in the greenwood

Wildwood Acoustic,
Slough Farm
Docker
Kendal,
Cumbria

First published in 2020
by Wildwood Acoustic
(Lloyd Music Ltd)

Bill Lloyd
Wildwood Acoustic,
Docker
Kendal,
Cumbria

Printed and bound in Great Britain

ISBN 978-0-9535238-2-5

Acknowledgements are due to the friends, family and colleagues without whom this book would not have seen the light of day.
Walter Lloyd for pointing the way.
Edward Acland for the Foreword.
Ali Lloyd for photographs, memory bank, proof-reading, good grammar and patience.
David Smith at *Indent*
for typesetting and design
Special thanks to Wallace Heim for reading the first draft by a novice writer and invaluable editorial guidance.

Images:
Sailmaker's Palm By Elgewen - Own work, CC0, https://commons.wikimedia.org/w/index.php?curid=13895926
St. Croix Wood - copyright unknown
Holme Wood By Jerry Evanu, CC BY-SA 2.0, https://commons.wikimedia.org/w/index.php?curid=12577885
Misty Forest by Artem Kavalerov
All other images unless otherwise attributed © The author.

Thanks to the people who inspired and helped me along the way, and who continue to keep the flame alive, and for whom the life of the woods is not an impossible dream but a daily reality.
I hope they all know who they are.

Cover illustration:

Penny Rock Wood, Grasmere, 1979

This edition is limited to 230 copies of which this is No. 195

I went to the woods because I wished to live deliberately, to front only the essential facts of life, and see if I could not learn what it had to teach, and not, when I came to die, discover that I had not lived.

(Henry David Thoreau, Walden, 1854)

We are as gods and might as well get good at it. So far, remotely done power and glory - as via government, big business, formal education, church - has succeeded to the point where gross defects obscure actual gains. In response to this dilemma a realm of intimate, personal power is developing - power of the individual to conduct his own education, find his own inspiration - shape his own environment, and share his adventure with whoever is interested.

(The Whole Earth Catalog – Introduction 1971)

Until you make the unconscious conscious, it will direct your life and you will call it fate. Your visions will become clear only when you can look into your own heart. Who looks outside, dreams; who looks inside, awakes. You are what you do.

(C.G. Jung, 1962)

Until one is committed, there is hesitancy, the chance to draw back, always ineffectiveness. Concerning all acts of initiative (and creation) there is one elementary truth, the ignorance of which kills countless ideas and splendid plans: that the moment one definitely commits oneself, then Providence moves too. All sorts of things occur to help one that would never otherwise have occurred. A whole stream of events issues from that decision, raising in one's favour all manner of unforeseen incidents and meetings and material assistance, which no man could have dreamed would have come his way. Whatever you can do, or dream you can do, begin it. Boldness has genius, power, and magic in it. Begin it now.

(W. H. Murray, 1951, quoting Goethe.[1])

See the whole thing is a world full of rucksack wanderers,
Dharma Bums refusing to subscribe to the general demand
that they consume production and therefore have to work
for the privilege of consuming all that crap they didn't really want
anyway such as refrigerators, TV sets, cars, and general junk you finally
always see a week later in the garbage anyway all of them imprisoned in
a system of work, produce, consume, work, produce, consume.
I see a vision of a great rucksack revolution thousands or even millions of young Americans wandering around with rucksacks, going up to mountains to pray, making children laugh and old men glad, making young girls happy and old girls happier, all of 'em Zen Lunatics who go about writing poems that happen to appear in their heads for no reason and also by being kind and also by strange unexpected acts keep giving visions of eternal freedom to everybody and to all living creatures.

(Jack Kerouac, The Dharma Bums. 1958)

1 *Translated by John Anster 1835*

Contents

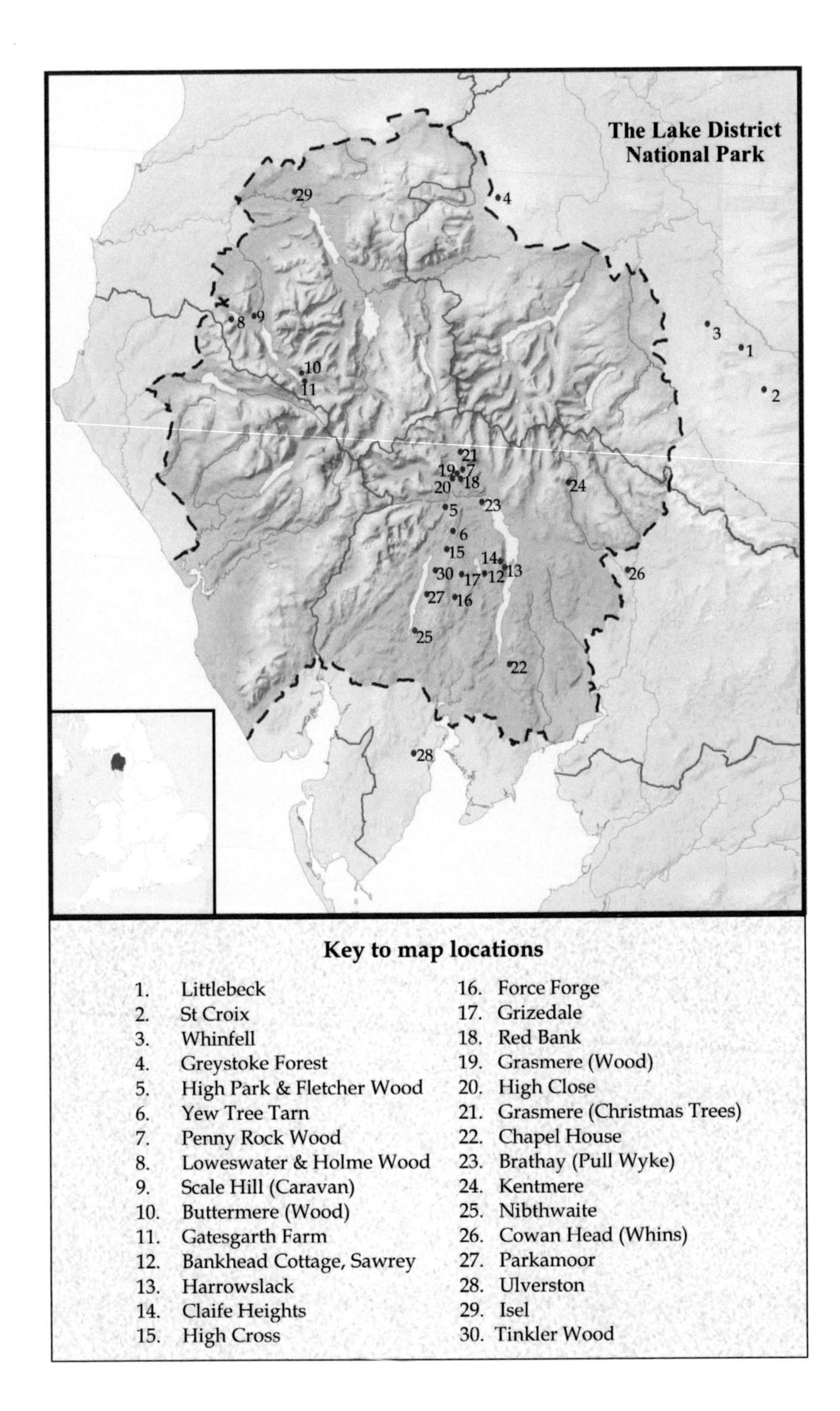

Key to map locations

1. Littlebeck
2. St Croix
3. Whinfell
4. Greystoke Forest
5. High Park & Fletcher Wood
6. Yew Tree Tarn
7. Penny Rock Wood
8. Loweswater & Holme Wood
9. Scale Hill (Caravan)
10. Buttermere (Wood)
11. Gatesgarth Farm
12. Bankhead Cottage, Sawrey
13. Harrowslack
14. Claife Heights
15. High Cross
16. Force Forge
17. Grizedale
18. Red Bank
19. Grasmere (Wood)
20. High Close
21. Grasmere (Christmas Trees)
22. Chapel House
23. Brathay (Pull Wyke)
24. Kentmere
25. Nibthwaite
26. Cowan Head (Whins)
27. Parkamoor
28. Ulverston
29. Isel
30. Tinkler Wood

Foreword
by Edward Acland

'One Horsepower' represents a single autobiographical chapter of Bill Lloyd's extraordinary life, with a special emphasis on his experience of horse logging in the Westmorland Lake District. Readers may well find sufficient guidance within these pages to take forward their own dream of working creatively with horses. This will involve managing such animals with love and understanding *and* as a legitimate way of carrying out difficult physical logging extraction tasks, often in very awkward terrain, inaccessible to fossil-fuelled powered machines.

Bill's passion in life, like my own, is firmly linked to and driven by a general anxiety relating to humankind's (human-unkind's?!) one-way trip exploitation of the bruised, abused and battered planet called Earth, which struggles to maintain living equilibrium within the delicate biosphere fabric within which we all live.

Bill's story is one person's description of a life-style that recognizes that a healthy planet can only be maintained by living in close harmony and within the limits of that body to support us. I have known Bill and Ali for many years. Our individual life-styles are different in many ways but tread the same basic path. All of us share a genuine belief that *homo sapiens (wise!?)* will need to let go of the collective fixation of aggressive consumer culture and find a finite resource forward plan, if we are to hang on to what's left!

I am convinced that a new crusade is taking place, and a new design for how we drive the planet is unfolding. It just needs to snow-ball, like a positive contagion, until every citizen comes to realize they need to join this new and exciting club.

Bill's story of his seven years' snigging timber has acted like a trigger to help fire up people following his good example. His story details the agonies, the grime, and the disappointments of his personal campaign, but above all else it reveals, via true grit, the humour, the dedication, the skill, and the special personal delight of being connected to the new age of enlightenment and gentle custodianship.

Edward Acland
Sprint Mill, Burneside

1. Hebden Bridge - the way in

If I had not cooked dinner for the Chairman of Lloyd's of London that evening at Lumb Bank, Ted Hughes' old house in Heptonstall, then I would never have started horse-logging in the Lake District. I should say straight away that the Chairman of Lloyd's was no relation of mine - there may be a connection between my family surname and the name of the world's oldest and best known insurance broker, but I am not aware of it. The Chairman in question was not even called Lloyd - he was called Havelock Hudson. A year later he was to become Sir Havelock Hudson, but when he sat down to eat that dinner he was just plain Hal.

The reasons why I cooked his dinner that night in 1976 are not directly relevant to this story, which is about horse logging, but they provide the background for the after-dinner conversation which sowed the seed of my decision to abandon any conventional career path and instead to follow an idea and an ideology.

After leaving Birmingham University in 1973 with a degree in Drama and Theatre Arts, I headed for London as aspiring actors do, and managed to get a job working for a pittance as an actor, musician and stage manager. I was lucky to get the job, but even so I decided that

London on the breadline was miserable and frustrating. The Actors' Equity minimum wage paid my rent for a small room in the outer reaches of Acton, and paid my tube fare to work so that I could spend two hours a day listening to the screeching steel rails, and it allowed me one good meal a day in the cheapest spaghetti house I could find. Although I had trained in the theatre and had come to London to seek my fortune, I realised I was not going to find it that way. I could get more satisfaction (and coin) by playing the tin whistle for cinema queues in Leicester Square for two hours than I could by working in community theatre for a whole week. The roar of the crowd and the smell of the greasepaint had drawn me to London, but it always seemed to be just round the next corner, so my partner Ali and I abandoned the city and retreated north, to my dad's farm in East Lancashire. Back where I started.

My dad, Walter Lloyd, mowing with Maggie, Whitworth, 1958

The farm comprised 80 acres of marginal land and shale heaps, 1000 feet up in the heavily industrialised Whitworth valley, in the Pennines near Rochdale. From the bedroom window where I grew up I could count 43 tall factory chimneys and see countless spoil heaps from abandoned coal mines and the vast quarries, which had provided the stone to pave

Trafalgar Square and build the waterfront quays in New York. In spite of the industrial landscape, it was nevertheless a farm, my home where my mum and dad had raised cattle, sheep, pigs, Fell ponies, and geese. My first memory of heavy horses is of sitting on the back of Fanny, our neighbour Billy Earnshaw's working mare, with my sister in front and my brother behind. My first memory of working a horse myself is of pulling a chain harrow when I was about 10 years old, sitting on the back of one of our own farm horses, Maggie.

My dad, Walter Lloyd, had graduated in Agriculture from Downing College, Cambridge in 1948, and started farming in East Lancashire, where his family had farms for four generations. Walter was a moderniser, and soon he brought the first tractor into the valley, a grey Ferguson, then later a Fordson Major, and before long he had given up his heavy horses. In the mid-sixties he bought a silorator (as the early forage harvesters were then called) and a trailer with a moving bed and so introduced silage to the Whitworth valley. The silage machines did not last – the cost was too high and profit too low and the ground was too wet, so it rusted in the field for 50 weeks of the year for the sake of two weeks' mechanised silage making. It soon became clear that the cost of the capital locked up in his rusting machines was eating up his slender profit margin, so he gave up on his small dairy herd and his pigs and changed direction. In 1959, after a memorable trip to see Fell pony breeder Sarge Noble near Shap in Cumbria, he introduced a herd of Fell ponies to East Lancashire, followed by a herd of Welsh Black cattle, both of which breeds are now dotted all over the hills and valleys. Apart from his wild ponies and his wild cattle, he is best remembered in Whitworth for his wild parties, which became famous in the most unlikely circles.

My mum and dad were helpful and friendly towards Gypsies and Travellers. They would offer stopping places to park trailers, along with grazing for horses and help with legal cases and planning applications. My mum, Vivienne, helped to teach Gypsy and Traveller children to read. A few Traveller families would regularly come and go from a piece of common land just outside the moor gate next to the farm known as 'Bacup Top' or sometimes just 'The Moon.' I would look forward to their

Bacup Top 1960.
Bill Lloyd,(left), Young Joe Cannon, and the Barkers, on 'The Moon'.

arrival, as it meant I could go out tatting with the boys, and at the age of 10 or 11, I first sat on a flat cart behind a pony and trotted up the highway to Bacup, and I did my first deal – for a pair of rusty hames.

So when Ali and I abandoned London in the spring of 1974, we headed for the farm with hopes of a more sensible life. We did a few repairs around the place, dug in about 400 yards of water pipe and helped with the horses and cattle, but we did not want to abandon our theatre training completely, so we decided to clear out the dark and dingy barn and turn it in to a performance venue. We cleaned the floor, lime washed the walls, built a stage, and hosted the very first practice meeting of the newly formed Whitworth Morris Men; we each had a pair of duck-toed dancing clogs made to measure by the clogger from Waterfoot. We rehearsed short plays by Henry Livings and learned the Pace Egg Mummers Play, which we performed up and down the valley.

It was a lively and formative time. My dad Walter organised an Alternative Technology weekend and 60 people came from all over the country – hippies, musicians, academics, environmental campaigners and political activists – to build three sorts of windmill, a methane digester, (how blind we were..) and a waterwheel which charged a 12v battery. We had a charcoal burn in a 40 gallon oil drum, epic conversations and plenty of home brew and scrumpy cider. On the Saturday afternoon, we milked a mare and hitched one of the cows to a plough. I built a *hangi* in the garden - a Maori cooking pit made of hot stones and fresh greenery, into which we carefully placed all manner of epicurean delights. That evening we were entertained in the newly converted Barn Theatre by a steel band from Moss Side, and the smell of *ganja* reached the main road half a mile away. Next day Walter built a yurt out of bamboo, and we welcomed the Krishna Temple, dined on their wonderful simple wholefoods and joined their ecstatic chanting for a blissful 6 hours that seemed like eternity.

After a memorable summer was over, it was clear that making a living in the theatre was going to be even harder in the Lancashire hills than it had been in London. We needed employment and we started asking around. Before long we had a call from my cousin David Pease, an Arts Administrator who had been a mentor to me while I was at University. He was now director of the Arvon Foundation, and needed caretakers at Lumb Bank, a mill-owner's house near Heptonstall. The house belonged to the poet Ted Hughes, later to become Poet Laureate, who had leased it to the Arvon Foundation to be converted into a residential school for creative writers.

My dad with Hades Hill Sally, 1963. The sledge brought in our supplies.

David offered us the caretakers' job of minding in the house, which had no electricity, no water, no bathroom, no sanitation, and no heating. We were required to provide security against squatters (Hebden Bridge was Squatting Central at the time) and to supervise the conversion of the house into a 20 bed residential centre. We fell in love with Lumb Bank as soon as we saw it - in a beautiful isolated location in the hidden Colden Valley, full of character, and with fine 18th Century cattle-sheds, stables, hay barns, stone water troughs and cobbled yards. It had not one but two superb walled gardens, with arched stone bee boles. It was a natural move for us - since the barn was to be converted into a theatre it was an extension of what we were already doing - combining farming with theatre. We accepted the offer, and moved over the hill into Yorkshire, carrying our entire possessions in the back of my dad's Ford Cortina.

Lumb Bank had a few acres of very steep but good quality grazing, so we took with us a small herd of Welsh Black cattle, which comprised three in-calf heifers, (recently bought in Dolgellau by Walter, about two weeks before he took off on a visit to Trinidad), and a Fell pony gelding. Within a week we had cleared out the cattle stalls and stables and started to break in the pony by persuading him to take a bridle, bit and saddle. We acquired a flat cart and a cat, and planted a tiny garden, growing vegetables for the table and horse beans for the pony. The crop of beans was so small we ate them ourselves like broad beans, and the pony had to do without, but we were getting a feel for self-sufficiency.

Over the next eight months the contractors came and went, the water and electric supply was installed, and the septic tank was built. The squatters in the cottages were evicted (but remained our friends), and the outbuildings were converted into bijou dwellings, offices and a print-shop. We became Centre Managers, and within a year we were elevated to Centre Directors, responsible for arranging, administering and supervising 40 courses a year, tutored by poets, novelists, photographers, musicians and playwrights. The work was extraordinarily intense, due to the hothouse environment created by a carefully constructed daily programme and the volatile combination of powerful tutors and impressionable mature students. It demanded long hours, creative and

Lumb Bank, Heptonstall, 1975

imaginative energy and well above average consumption of alcohol, to say nothing of managing the internal and external politics and the constant financial pressure.

Although the Arts Council Literary Panel was generous, (if occasionally snooty) Arvon was always short of money, but luckily we had heavy friends. Ted Hughes himself had mythic national status as a writer, and he put his weight behind the project. At some point, one of the friends introduced Havelock Hudson, recently elected Chairman of Lloyd's of London, Insurance Brokers. After Ted Hughes, Hal was without doubt our heaviest friend. He donated generously and introduced us to other heavy supporters. From time to time he would travel from London to visit the new centre, and talk finance with the Council of Management.

Part of our job as Centre Directors was cooking supper for participants on the first night of every five-day course, and we had a splendid new kitchen and rustic dining room to seat 20. On this particular evening in early 1976, we cooked dinner for the Council of Management, the staff at Lumb Bank and Totleigh Barton (our sister centre in Devon) and for our honoured guest Havelock Hudson.

After dinner, settled in the Conran armchairs, with full wine glasses and cigars burning nicely, the conversation turned to politics, economics, and finance. The country was in crisis at the time. Within the previous couple of years Britain had endured an oil crisis and the three-day week. *A Blueprint for Survival* had revealed the dangers of environmental catastrophe. The depletion of the ozone layer by CFCs had just been proved. The number of violent deaths in Northern Ireland had passed 1000, unemployment and inflation were at exceptional levels, and there was an enormous balance of payments deficit. The government was battling with the Trade Unions, the Labour Party was tearing itself apart, and Sterling had collapsed. When Cabinet papers were released 30 years later, it turned out that at the very time we were making dinner for Hal Hudson, the Government was negotiating an enormous and unprecedented loan from the International Monetary Fund. In return for bailing out the UK, the IMF demanded deep cuts in public expenditure which were to have profound effects on economic and social policy.

Suddenly we were in after-dinner conversation with Hal Hudson, who was Chairman of the company which employed the country's top actuaries. He dined with the Prime Minister and had the ear of some of the most powerful movers and shakers in the land. The conversation that followed planted itself in my mind and influenced my thinking on politics, economics and the environment for the next 35 years. Hudson was necessarily discreet, as he knew as well as anyone that our currency, and with it our whole financial system, depended on 'fiat money' which in turn depended on confidence. His firm had to assess the scale, the nature and the effects of the risks associated with different policies and different parties, and he knew the score. He had to be very careful what he said. He could not risk undermining confidence in the resilience of the financial system by spreading alarm, but nor could he pretend that there was no problem, since that would make him look either foolish or dishonest, and he was neither.

When I asked him courteously about his view of the financial crisis from his elevated position, he would not discuss any specifics, but he was very clear that unless a solution to the crisis was found within 6 weeks, the country would undergo a profound constitutional, financial and probably social collapse, on a scale never seen before. We would be broke. We were on the edge of catastrophe.

In 1976, in the days before Google, and before every point of view, no matter how apocalyptic, came to be represented on the Internet, Hal Hudson's analysis was strong medicine. My natural cynicism and leftish anarcho-hippy tendencies made me deeply suspicious of mass-media received opinion, whether from the left or right, but now I had it from the horse's mouth: our entire state, our whole culture, was in reality not solid and dependable, but fragile and deeply vulnerable.

Even now I ask myself why Hal Hudson's bleak vision had such an effect on me, but how could I *not* believe this man? The country was clearly in deep crisis, and his firm was in the front line of trying to sort it out. Why would he lie to us? He was our friend, our supporter; his generosity probably paid my meagre wages, and I had just cooked him dinner. He was charismatic. Maybe the charisma came from his wealth and power, but that was not the whole story - he was a decorated soldier, he drove a Bentley, he wore tailored clothes like none I had ever seen, he spoke with authority, and he was the grandest man who had ever stooped to have a proper conversation with me about the things that really mattered. So I listened carefully, and remembered what he said.

At that time, the things that mattered most to me politically were the fragility of the global environment and the need for a new social, political and economic model that would not bring us all to catastrophe. In 1974, after my dad's alternative technology weekend, I had read *'A Blueprint for Survival'* which, along with the *'Whole Earth Catalog'* was a key text for the alternative society of which I already felt I was a part. Something had to be done. As the cover of the *Blueprint* proclaimed *"After reading it, nothing seems quite the same any more."* The *Blueprint* argued for a radical restructuring in order to prevent what the authors referred to as *"the breakdown of society and the irreversible disruption of the life-support systems on this planet."*

My reading provided well prepared ground, Havelock Hudson sowed a seed and my instinctive ideology watered it in. Hudson confirmed the validity of my deep suspicions about the nature of the state, my fears about the fragility of our materialistic culture, and the enormity of the confidence trick that had been perpetrated on the people. My lefty-leaning university education had convinced me that consumerism, nurtured by the drip feed of mass-media advertising, was an essential component of capitalist growth. The defenders of capitalism and consumerism argued that despite its defects, the free market was morally superior to totalitarian states, and brought a higher standard of living. I suspected otherwise, but had no evidence until Hal Hudson's bleak predictions convinced me that my suspicions were not mere left-wing paranoid obsessions, but were echoed calmly and rationally by the highest and mightiest in the land, using legitimate arguments based on actuarial and empirical criteria. If the apparently stable social fabric was in reality vulnerable, fragile and precarious, then the justification for an environmentally damaging oil-based consumer economy was false.

My alarm about the dangers of ecological and economic disaster merged with inspiration from great writers, in particular Ted Hughes, Basil Bunting and the Beat Poets, Jack Kerouac and Gary Snyder. At Arvon I was surrounded by writers, but it was clear to me that we needed more than words – it was time for action.

Fell pony breeder Sarge Noble riding a Heltondale pony near Shap, 1959

2. Pentre Evan - ideology

The Arvon Foundation was an extraordinary and inspiring immersion into poetry, politics, community living and group dynamics, but after two years Ali and I were mentally, emotionally, physically and financially exhausted.

I worked 70 hour weeks arranging and running the programme of 40 courses per year while Ali worked as a teacher in a comprehensive school on the other side of the Pennines, and with no wheels of her own she had to rely on getting lifts or sketchy public transport. At Lumb Bank we were caught between two different sets of millstones. On one hand there was an unending feud between three factions of the poetry world. Modernist poets Jeff Nuttall, Eric Mottram, and Tom Pickard were battling with the Poetry Society, the Arts Council Literature Panel and the Arvon Foundation Council of Management, and we were implicated. On the other hand was a disagreement between George and Christine Tardios (Centre Directors at Totleigh Barton, our sister centre in Devon) and the Arvon management over the role of the tutors and the structure of the courses. In both struggles we were required to take one side or the other, which was difficult. We had some sympathy with the work of the modernist poets, who complained about the exclusion of anyone except cosy 'drawing room' poets, but it was their opponents on the Arts Council panel who paid our bills and called the tune. George Tardios was an original, all-or-nothing, 'either-it's-black-or-it's-white' Greek Cypriot Londoner, for whom using his fists might be preferable to fence-sitting. The Arvon Foundation management was led by my cousin and mentor David Pease, to whom I owed a debt of loyalty and who had given me the great opportunity of running the Centre. I took sides mostly with Tardios, who by that time had become a friend, but the consequent stress of managing the radical poets, keeping the Arts Council happy and avoiding a family falling-out was almost unbearable, and in our exhausted state it was unsustainable. So we left.

Bill and Ali on the Flat Cart at Lumb Bank 1975

The Welsh cattle and the Fell pony went back over the hill to Rossendale, and we took off. I had a few solo adventures, including a skirmish with an ex-Korean War pilot, Taylor Collings, who owned part of the Blasket Islands and who employed me as 1st Mate and unwitting front man for a dodgy deal involving an 80 foot boat on the Cote d'Azur. I managed to avoid jail and get home, but that is another story. It was the hot summer of 1976; we were both 25 years old, full of energy and ideals and after spending two inspiring years with some of the finest poets and novelists in the English language, we were ready for anything.

Before long we found ourselves in Pembrokeshire, on a farm called Pentre Evan (or Pentre Ifan) home of Satish Kumar, editor of *Resurgence* magazine. A small, alternative, low impact community had taken root, and our friends Tony Ashford and Viv Lewis were living there. Tony and Viv had met my dad Walter in 1973 at the Trentishoe Whole Earth Fair, where they were baking wholemeal bread in a home-made oven. Walter bought a loaf of bread, struck up a conversation, and before long Tony and Viv landed at the farm in Rossendale. They brought with them exotic goods - bulgur wheat, tamari soy sauce, seaweed, miso, apple juice, tahini, wooden eating bowls, and a pet rabbit. (Strange as it may seem, we had never seen any of these things before, but we soon become devotees of wholefoods.) As we moved into Lumb Bank, they moved to Bricket Wood in Hertfordshire and two years later, as we moved on, so did they. We were all looking for new directions and 'alternative society' beckoned. They headed for the community at Pentre Evan, so we decided to check it out, hitch-hiked through Wales and moved in to the farm with nothing but two rucksacks and a sense of excitement and discovery. The ethos was part agricultural, part political, part spiritual, with a deeply idealistic dedication to sustainable living. We tended the vegetables, milked the goats, split the firewood, ground wheat into flour and made our own bread. Nearby was Fachongle Isaf, home of John Seymour, the great self-sufficiency guru of the day, who wrote one of the key texts of 1970s hippiedom, *The Fat of the Land.* He put his ideas into practice, running a small farm helped by dozens of aspiring and idealistic twenty-somethings who, like us, wanted a piece of the new dawn.

Fachongle Isaf, and much of that part of Pembrokeshire, was full of like-minded young idealists, for whom the important markers of identity and status were a Morris 1000 van, or a chillum, and preferably both. The first time I visited Fachongle, the foreman (a woman) was enveloped in smoke, using a red hot poker to make chillums for sale at the forthcoming free festival. (Now that's what I call capitalism.) It was at Pentre Evan that I was first encouraged to seek enlightenment by sitting on top of the cromlech at the full moon and drinking mushroom tea. (It was not until nearly 20 years later that I got to know Mr. Shroom, in the woods at Crooks Farm, near Bouth, with my brother Tom and a posse of Peace Convoy yurters, but that is yet another story.)

It was at Pentre Evan that for the first time I yoked a heavy horse, Beauty, and dragged a log out of a wood, not knowing that would become my life for seven years. At the Pembrokeshire summer festival, Megan Fair, there were more firsts – the first time I volunteered to construct the toilets for a hippy festival (how idealistic is that?), the first time I smoked a fish in the chimney of a wood-burner (I didn't inhale), and the first time I sat with friends singing songs round the fire through a long summer night, and watched the sunrise bring that glorious and unmistakable sense of renewal, community and shared beauty. Pentre Evan was a meeting of different ways, different paths. Thanks to Satish Kumar's ideals, John Seymour's practicality, and the hedonism of the hippy generation, we acquired the habit of alternative living, the confidence of youthful idealism, and the beginnings of the practical skills to combine them. We embraced those three different paths with equal enthusiasm.

Pentre Evan was too pure for the likes of us, but it was formative and highly influential. The summer of 1976 made me realise that I would probably not be happy in a suit, working 9 to 5 for a pension, or even going back to arts administration. Although the emotional sophistication of the theatre and the literary feasting of the Arvon Foundation were tempting, they seemed too cosy and domestic, too ephemeral, too mundane. I was hungry for something more fundamental, more effective, more definite, more mythic and more heroic.

The gurus of the early green movement such as Seymour and Kumar attracted a potent following, and that brief but formative summer had connected me firmly with a network of people who shared something of that hunger. Tony Ashford and Viv Lewis went on to establish their own dairy herd of Jersey cows, but not before they had introduced us to whole foods, home baking, and to a network of like minds. Pete and Mel Landells taught us about agricultural economics, home brewing, the absolute necessity for focus on the job in hand, and true cynicism.

Preseli Kitchen, Megan Fair, Pembroke, 1976.
Bill heats up a bodhran drum skin over the chimney,
Ali stirs the wok. Lewis rolls a smoke with Jamie behind.

Lewis Cleverdon became a wheelwright in Pembrokeshire and 40 years later is a hill farmer and eco-commentator. (It was Lewis who suggested touring the country pubs of England to find quality horse harness hanging on the walls, to buy it back in case the oil ran out.) Rick became a shepherd in The Lake District and in Galloway, and died too young; George Van Wienen became a silversmith in Manchester before he too died, tragically killed on his motor-cycle; fair haired Jamie taught us all about ecstatic dance and mushrooms, and then vanished; they and many more whose names we have forgotten, but whose faces recur in our dreams, represented a new movement, a new tribe, a new family.

The ideology at Pentre Evan was extreme – righteous, laudable, but out of our reach. It was best illustrated one morning at breakfast when it was announced that the barley crop (80 acres) was ready to harvest. No combine-harvester was available, and anyway Satish had already suggested that we should harvest it ourselves, by hand, with scythes and sickles, as they did in India. Eyebrows were raised, doubts were muttered, but we set out on foot for the far fields, with our hand tools and a few bags of bread, goat's cheese and lettuce. We laboured all day and returned in the evening exhausted, blistered and hungry, with just half an acre of barley down, bound into loose sheaves, tied with straw rope and standing in tatty stooks.

We were not disillusioned, because we had never been convinced. Satish Kumar was another charismatic, and by his simple faith and his example had persuaded us to try to harvest 80 acres of barley with sickles, even though our rational, conventional, cynical minds knew that it was hopeless. We knew well enough that for the past three thousand years people had harvested grain, by hand, with sickles more primitive than the ones we used, and so we knew that it was not only possible, but commonplace. We also knew that we could not do it, because we did not have the skills, the stamina, the time or the motivation born of necessity. We were too few and we had easier options. Possibly Satish thought we were too lazy and too decadent, but he never said so. We had tried and failed, but that did not make his aspirations unrealistic, or diminish his ideals. Failure is not unusual. As my dad Walter told me, 'Experience is the name we give to our mistakes, and I have a lot of experience.'

At Pentre Evan with Satish I began to wonder how it happens that humans, so logical in other ways, will try to do the impossible, for the sake of an abstract or spiritual ideal. Satish had been a Jain monk. He had travelled with Vinoba Bhave, considered to be the spiritual successor to Mahatma Gandhi, and inspired by him he left Delhi and walked barefoot to the capitals of the big four nuclear-armed countries - Washington, London, Paris and Moscow, on a peace-walk of over 8,000 miles. His reverence for nature was at the heart of every political and social debate in which he engaged. He was an idealist, not a realist, so for him a small group of novices harvesting 80 acres barley with sickles was more than a gesture - it was the embodiment of an ideal. As Satish put it: "Look at what realists have done for us. They have led us to war and climate change, poverty on an unimaginable scale, and wholesale ecological destruction. Half of humanity goes to bed hungry because of all the realistic leaders in the world. I tell people who call me 'unrealistic' to show me what their realism has done. Realism is an outdated, overplayed and wholly exaggerated concept" [1]

In Kumar's philosophy, humanity should not be limited by what our conscious minds consider to be 'realistic.' Instead, by over-reaching ourselves, by striving for the impossible, even in the full knowledgeof the difficulty, we might sometimes achieve it. The unexpected might happen for many reasons - but if we think we cannot do a thing because we believe it to be impossible, then we never try and so it becomes impossible, but only because our minds make it so. It takes courage or blind faith to do something which is considered impossible. Someone, often a charismatic, will attempt the impossible and succeed, and instantly the impossible becomes feasible. No-one would have believed that a solo ascent of the North Face of the Eiger, or a wing-suit jump from its summit, was a realistic possibility, until it was done.

When I describe Satish Kumar, or Hal Hudson, or Ted Hughes, or Jack Kerouac as charismatic, I am not referring to godheads or divine inspiration. Charisma does not rely on wealth, or even good character, but to the fact that others, usually subconsciously and for different reasons, are attracted to their manner, their conduct, their beliefs, or their achievement. Because they can do things which others cannot, people find themselves respectful, loyal, slightly in awe. They will treat them as leaders and follow them into the new territory which they have discovered, explored, and charted. Charismatics can be dangerous of course - they can inspire criminal gangs, fundamentalist cults and mass suicides. Their talent for convincing the gullible can, and does, lead to delusions of world domination, but the blind loyalty they inspire can also lift the spirit, and provide beacons of hope in dark and gloomy minds.

In short we did not quite know why we followed Satish Kumar for a few miles on his hard road, but we knew that we made the right

1 *Sica, Giulio (2008-01-16). "What part does spirituality play in the green movement?". The Guardian (London). http://blogs.guardian.co.uk/ethicalliving/2008/01/what_part_does_spirituality_pl.html. Taken from WIKIPEDIA*

decision to follow him, and we made the right choice to leave him. Satish was an inspiration, who taught that idealism and spirituality could give our lives meaning, and he was also the catalyst for our determination to take action, to do something solid, concrete and tangible.

We may have rejected some of his ideas as impractical, but we did not abandon ideology, which gave us our direction and motivation. Instead we simply developed our own ideology. The Arvon Foundation had opened my eyes to a new landscape of literature, and at Pentre Evan I identified with the philosophical writings of Henry David Thoreau, Jack Kerouac and the *Whole Earth Catalogue* to such a degree that they became semi-sacred texts. The reverence for nature in William Cobbett's *Cottage Economy* and George Ewart Evans' *The Horse in the Furrow* was no less powerful than the aspirations of the *Jain Dharma* advocated by Satish Kumar, but it was closer to our own temperaments. Our heroes and guides were practical farmers and craftsmen who could make something out of nothing, working with natural forces. Geraint Jenkins' *The English Farm Wagon*, and George Sturt's *The Wheelwright's Shop* not only provided technical detail about designs and methods, but they embraced the social and communal aspects of village life, the diversity of folk culture, and the possibilities of living a life rich in meaning and satisfaction, not based on conspicuous consumption and with our feet on the ground.

The village wheelwrights, harness makers and horsemen achieved a peak of sustainable craftsmanship, before the discovery of plastics and the internal combustion engine, and before mechanised mass production began the industrial-scale exploitation of the planet. They combined low-impact and organic sustainable technology with fine aesthetics, functional design and high craft skills. They used local wood, locally wrought iron and simple tools in their bare hands to produce beautiful and sophisticated road vehicles and elaborate agricultural machines, all powered by horses. Fascinated by their insights, I devoured every book I could find about wheelwrights, and seriously considered training in that craft. I chose to become a horseman instead, because I was betting that the oil would soon run out and that the whole petro-chemical fuelled machine would then grind to a halt. Horses offered an immediate solution – instant traction.

I was convinced that horses were the way of the future, as much as the story of the past. I was wrong then, of course, and stayed wrong for 35 years, although the warning lights are still glowing brightly as the burning of fossil fuels shows no sign of slowing and the ice melts..

A Blueprint for Survival and *The Limits to Growth* had given us evidence that economic and ecological catastrophe was around the corner, and Hal Hudson's insight confirmed that our financial system was fragile and approaching a cliff-edge. The classic books about English rural life before the industrial revolution showed us what we had lost, and we traced a direct line from those writers to the American Beat Poets and from there to the imperative for self-sufficiency realised by John Seymour and the *Whole Earth Catalog*. The road ahead was clearly marked. Satish Kumar's example, through his uncompromising idealism, convinced us that our aspirations were modest and realistic by comparison.

We could not sustain ourselves at Pentre Evan except at subsistence level, and we did not want to become monks, nuns or hermits, so we could not stay to support Satish in his bold venture, but our introduction to alternative culture showed us that we were not alone, and it was there that we formed our plan. Although we were convinced and motivated, we recognised that the ideology of our new-found tribe was out of step with the rest of the world. We were part of a small minority, whose strong ideals were derided by all 'right-thinking' consumers as 'crazy hippies', 'brown ricers', 'yoghurt weavers', or lefty radicals. Although we believed that consumerism and materialism was a social, moral, spiritual and environmental blind alley, and the search for an alternative seemed to be plain common sense, we were still regarded as hopelessly idealistic and backward-looking by the majority. Pentre Evan and Satish Kumar gave us a way out – we could not honestly embrace his spirituality, but he showed us that living for a social and environmental ideal was an honourable path, and one that had some hope of success if it remained grounded and practical. With enough like-minded people to support each other, and a strong enough belief in what we were doing, we might try to change the world. We decided we would try to do it with heavy horses.

3. East Yorkshire - heavy horses

In the 1970s, energy was the focus of environmental politics. The oil crisis, the miners' strike, the three day week and power cuts, and the dangers of nuclear power and its residues, (culminating in the Three Mile Island meltdown in 1979) all pointed to the conclusion that the growth of industrial capitalism and consumerism was limited by the planet's resources of energy. *The Limits to Growth* (1972) had confirmed the forecast. We did not worry about climate change in those days – it was as yet hardly even identified as a problem, although CFC damage to the ozone layer had been confirmed. We believed that the oil reserves would simply run out, and soon. No viable alternative energy sources had yet been embraced by government, solar panels were still at the laboratory stage, and the only windmills most people were aware of were on tourist pictures of Norfolk, Holland or the Greek islands.

We talked for hours about which alternative future was most likely to become reality when the lights went out. Like most environmental activists we believed that the internal combustion engine was the devil's agent (albeit a temporary necessary evil to allow our personal mobility, of course). We seriously discussed the time when motorways would be used only by horses, which could take advantage of thousands of acres of good grass on the vast green verges. We were sure that nuclear power would inevitably kill us all sooner or later, so wood fired steam engines would require thousands of acres of firewood. We were certain that landfill sites full of plastic mickey-mouse gadgets and broken but perfectly repairable domestic appliances were an affront to any species which dared to call itself rational.

It was clear that what the world needed was a sustainable, affordable, self-reproducing, controllable, beautiful, enjoyable, safe and friendly source of power and it was as plain as the sun in the sky that the world already had one - the working draught horse. The horse had been the primary source of traction available to humans for at least four thousand years. The horse had built Western civilisation, albeit slowly. Horses can live and work on grass if necessary, and they produce their own replacements. We were convinced that the horse had to be the way forward, and in 1976, for anyone in the UK interested in heavy horses, all roads led to Geoffrey and Lucy Morton and their children Janet, Andrew and Mark, and their farm at Hasholme Carr in East Yorkshire.

Geoffrey Morton, dragging stubble with an eight-horse team, 1976

The Mortons ran their 120 acre arable farm using only heavy draught horses - Shires, Clydesdales and a few Ardennes. Once again Tony Ashford and Viv Lewis gave us the contact, and we made a trip to East Yorkshire. The Mortons were not purists, and not particularly idealistic, but they were hard-nosed and exceptionally hard working farmers. They employed a contractor, using a machine, to spread slurry on the stubble from time to time, and they had a Landrover which they would occasionally use as a tractor to bring in an odd load of corn if necessary, but otherwise, they used just horses. Their operation was unique in the UK, and photogenic so attractive to TV companies, and they ran farm open days that brought in 1000 people at £5.00 per head, so they had other sources of income, but their horses provided the main traction.

They worked the farm more or less as the Victorians did it, because it was the best way for them. They ploughed, either with pairs or with a big team, they harrowed, they rolled, they drilled (planted), they weeded; they cut corn with a binder and hay with a grass reaper. They spent long days setting up the sheaves in stooks to ripen, then spreading them out to dry after a shower, then more days stooking them up again.

The Morton brothers binding corn with a four-horse team. 1975

When the sheaves had dried and ripened, they were loaded with long handled forks onto horse-drawn drays (known as lorries) which were led to the stackyard, to be stacked 20 feet high and beautifully thatched for the winter. When the steam-powered threshing machine arrived, a team of a dozen men using pitch-forks would feed the machine, then carry the corn on their backs in 16-stone (about 100 kilo) sacks up the steps into the granaries. They baled up the straw in a steam-powered wire-baler, working in a tight team with unforgettable *camaraderie.*

Geoff Morton was known to take on aspiring horsemen who were idealistic enough to work for nothing except their keep, and in return for their work he would show them the ropes and the techniques of working and managing heavy horses. After helping out at an Open Day and showing that I could work, I signed up. The arrangement was somewhere between an old fashioned apprenticeship and an Australian jackaroo. I had my own room in the farmhouse, and I ate and worked with the family. They ate prodigiously – enormous and frequent meals, all prepared by Mrs. Lucy Morton and their daughter Janet, who were both teachers when not in the farm kitchen. The men were all heavyweights, fit and strong - Mark Morton could lift two four-stone weights above his head simultaneously and was a champion arm-wrestler, immensely proud of his strength.

With the Mortons I learned about feeding, shoeing, harness, breeding, breaking, and horse management. I learned incidentally about ploughing and cultivation, harvesting, loading and unloading drays, and threshing and stacking corn. By the end of my six months apprenticeship I was trusted to drive a pair of Shire horses in a grass reaper and one fine summer day I cut a three acre field and Ali drove the tail-rake, pulled by Bill, the Ardennes stallion.

Best of all, by the end of my stay with the Mortons I had worked a complete cropping cycle; starting with ploughing and cultivating the land, then drilling the corn seed and watching it grow, harvesting, stooking, leading, stacking and threshing the corn. Once the corn was safe in the granaries, we transported a few sackfuls from East Yorkshire to the watermill at Little Salkeld, in the Eden Valley, where it was ground into flour to be made into bread at the Village Bakery in Melmerby.

The Salkeld Watermill and the Melmerby Village Bakery were both newly established, driven by the same holistic ideology that we had embraced. We had joined a new tribe, and passed another milestone - we could now eat fresh-baked wholemeal bread, made from wheat that we had personally sown and harvested using nothing but renewable energy - horse-power on the farm, water-power for the flour milling and brushwood for the wood-fired bake oven. This was the way forward.

Good wholemeal bread became a symbol and a touchstone, an essential first step in making our escape to a better way of life. Eating good bread, made from our own grain and our own labour, we could feel, at last, that we were escaping the infernal march of planetary destruction and becoming part of a self-sustaining wheel, in harmony with the good earth from which we came and to which we would return.

Ali, tail-raking with Bill the Ardennes stallion, in East Yorkshire.

Although this feeling was humbling, and we disdained the conventional marks of status which defined individual achievement - flash cars, fashionable clothes, conspicuous shopping, the mortgage and the credit card - there was in that apparent humility and disdain a real excitement and a kind of pride. By trying to save ourselves, and living the way we chose, we hoped to demonstrate that sustainable living was not just possible, but relatively simple to achieve. We were on a mission.

Thanks to Geoff Morton's single-minded determination to keep his heavy horses, we had found our gateway, a light to lead us out of the darkness. But I soon found out from Geoff that I had no prospect of making a living with horses by becoming a farmer. A farmer needs heavy money for land, and a great deal of skill. I simply did not have the heavy capital or the knowledge, and even Geoff, the most successful heavy horseman in the country, could not afford to pay me a living wage.

He suggested that the only way an aspiring horseman might earn a living at that time was in forestry. Horses were still in demand to pull timber out of commercial woodland, and he knew of two or three places where horsemen were still working in forestry - the Welsh Borders, the New Forest, and the North Yorkshire Moors. He gave us the name of a forest horseman in Dalby Forest in North Yorkshire, and we immediately knew that our next step on the mission was to go and find him.

Andrew and Mark Morton, binding with four-horse team. Hasholme Carr, East Yorkshire

4. Appleby Fair - on the road

We still had to earn a living in the meantime. Ali and I had moved back to Hebden Bridge, where we lived in a one-up one-down bottom terrace on Eiffel Street, Birchcliffe, with a shared outside toilet and a coal-chute from the street above directly into the coal-hole next to the kitchen. My theatre training was enough to get me a job as stage-manager at the Halifax Victoria Theatre. For months I cycled the 9 miles to Halifax and 9 miles back, often at 2 o'clock in the morning, to stage-manage a series of mainstream pop acts - David Essex, Cliff Richard, Freddie Starr, Alvin Stardust, and the Bay City Rollers. The spectrum of shows ran from Big Daddy (the All-in Wrestler, not the rapper) to the English National Opera, and back again. It was exciting enough, and demanding, and it paid the rent, but I was impatient to get out into the world and become a horseman.

My dad, Walter, encouraged us at every turn, and together we decided to go 'by road' (i.e. horsedrawn) to Appleby Fair, the greatest horsefair in Europe, and an annual pilgrimage for ten thousand English Gypsies and Travellers. At that time Walter was working as the Emergency Planning Officer for Greater Manchester Council and spent his days making plans for what to do if a chemical works exploded at Runcorn or a jetliner crashed onto the city centre. He had a week's holiday due, and Appleby Fair coincided with Ali's half term from her job as a pre-school visitor and I took a week off work. Walter swept out his bow-top wagon and invited his secretary and her partner to come along for the ride, although in fact they walked almost all of the way there and back. Our neighbour, David Westerman, fancied the trip with his coloured pony, and we thought that a spare horse and an extra set of hands would be helpful, so we agreed to take him along with us.

We made it from Whitworth to Appleby Fair in three days. Looking back, that was remarkable, as the breakages came one after another, the horses were unfit and we were a little overloaded. In later years the trip became an annual event, and we had learned to set off only with fit horses and well-fettled vehicles; by carrying only essentials we could halve the weight, we could ride instead of walking and trot almost all the way. It would still take three days, but more of each day could be spent resting and enjoying the life of the road instead of trouble-shooting.

Appleby Fair, with brother Tom and Ginger in the River Eden.

On our first trip, in 1977, Walter had to get back to County Hall and I had get back to the Victoria Theatre, so we had only three clear days on Appleby Fair Hill before we had to set off back again. We made it home in another three days, and on that short trip we had learned all about travelling 30 miles a day, pulling a heavy wagon under pressure. We found out about living with horses day and night, where and when to make camp, how to carry water, how to cook on an open fire, and where to find firewood in the hedgerows. (Old dead thorn trunks out of a hedge bottom are in a class of their own for heat and for cooking.)

At Appleby Fair we gained a suggestion of respect from some of the Gypsy people simply because we had travelled 'by road', at a time when there were only a handful of bow-tops making that trip. We were Gorgios (non-gypsies) and many Gypsies, especially the young boys, have disdain for Gorgios whoever they are – but particularly for hippies,

Ali at Appleby Fair

and we were clearly as close to hippies as made no difference. The fact that we had covered 90 miles in three days with green horses and untried wagons gave us a notch of credibility. A notch was better than nothing, and although Walter already had some respect among the Gypsies for his work in helping to save the Fair against the constant threats of closure, we had to earn it for ourselves. We had made a start, and made some friends.

At Appleby we re-established contact with Tony and Viv, who had left Pentre Evan in Pembroke and were by then living at King's Meaburn, not far from Appleby town, working for Gordon Jackson at Littlebeck Farm. Before setting off for home we yoked up the tub trap and drove out to see them. While in the queue of motor traffic waiting to turn right to cross the bridge over the Eden by The Sands in Appleby town, a Traffic Policeman suddenly remembered that horses were not allowed over the bridge, and threw up his *Dayglo* arm to stop us crossing. The pony was naturally startled, and took two steps backwards. The car behind was too close, and the iron step at the back of the tub-trap went through his radiator. The policeman escorted us back to the Station where the unfortunate car-driver naturally blamed me, and I blamed the lack of police and their lack of training with horses. Fortunately there were enough witnesses to the fact that the policeman had raised his hand suddenly and without warning, and the horse had only done what a horse will do. Realising that they might be digging a hole for themselves, the sergeant released us and we were sent on our way. All I could do for the driver was to suggest that he make a claim against the policeman, which seemed fair enough at the time, although he was less than happy and the sergeant gave me a hard stare. It was an unfortunate accident, but it could have been worse, and we put another lesson down to experience.

The five mile jaunt to Littlebeck was a joy. Lightly loaded, in no hurry, we trotted along the quiet back-road through idyllic Westmorland landscape to a big old farmhouse by the Lyvenet river. We came to love Westmorland farmhouses over the next 30 years, and we were entranced by the practical, functional beauty of the place. Low ceilings, a wood fired Rayburn, stone-flagged floors, mullioned windows with stone window-sills, and plenty of spare bedrooms – our dream house. Outside were workshops, ancient cattle stalls, granaries, a garden with apple trees and a shed full of firewood. Tony and Viv, ever generous, suggested that if we moved in with them, we could share the rent and their Morris 1000 van, help out on the farm, maybe do some baby-sitting, maybe find some forestry work, and keep our ear to the ground for timber-horse work. It seemed like another door opening, and of course we went through it. We talked it over on the long road back to Rossendale, and by the time we got home we had decided to quit our jobs, and make our move.

That was the big decision, the moment of truth, the 'Goethe' moment. We realised that we were ready to give up our regular wages to strike out into the unknown, with no certainty that we would find what we wanted. We knew that it would not come to us, so we had to go and find it for ourselves. Suddenly, almost overnight, our mission had begun.

With David Westerman an the road to Appleby Fair

I gave my notice to the Victoria Theatre, and just as soon as we could we were on the road again, on a road trip to find Geoff Morton's mysterious forestry horseman in the North Yorkshire Moors.

We arranged to borrow Walter's Fell pony stallion, Hades Hill Charlie, and fettled up the old Governess Cart (also known as a tub trap) which we had taken to Appleby. I spent a day oiling harness and packing tools and equipment into a tin trunk, with our bedding and feed for Charlie in fertiliser sacks. We greased the axles, fitted a new step, gave it a new coat of paint, then packed up a tent and a kitchen box, a tethering chain in a bucket, and a rucksack of clothes, just leaving room for us to sit down. Next morning we set out early to catch Charlie, who was running wild with his herd of 20 Fell pony mares on the moorland above the farm.

All round the edges of these moors were patches of green grass – an abandoned intake here and there, or good farmland sold to developers and grazed by all the commoners until the bulldozers moved in. We knew from experience that the herd was likely to have looked for the best grass, and we found them on a football field about five miles from home, and only about a mile from the town centre, grazing happily and certainly not wanting to be caught. We drove the whole herd back onto the common and headed them home along the winding moorland tracks. They knew where they were going and, led by the stallion, they soon set off at a canter, manes and tails streaming, and waited by the moor gate at Duckworth for us to catch up. Young Joe Cannon the farrier was waiting for us in the yard to clap on a set of new shoes, and in the calm of a summer afternoon we set out on our first solo trip. As the sun was setting over Pendle Hill we camped in a glade by a backroad, sniffed the woodsmoke and the sweet air, and started our great adventure. After a year of dreaming, reading, talking and planning, we had made our first move. Action at last.

We were away for two weeks, travelling past Pendle Hill, up the Ribble valley, skirting the Forest of Bowland to Settle, then up through Wharfedale, 'over the top', down Widdale and Wensleydale, across the plain of York, round the Hambleton Hills, and back via Otley, Baildon and Keighley. We camped and grazed the horse on verges and lay-bys, found shoeing smiths as we needed them and lived on porridge oats, bulgur wheat, bacon, cheese, almonds, apricots and chocolate. We experienced for the first time the steady slow progress of a pony and trap through the vast Turner-esque landscape on the plain of York, watching the rain squalls appear in the distance and trying to guess if they would hit us or pass us by. We became adept at pulling our waterproofs on and off quickly, and finding shelter from the worst of the wind and rain.

There is a peculiar and unique pleasure in the first five minutes of travelling horsedrawn after taking to the road in bright sunshine when a storm has passed - a sense of survival, renewal, and *catharsis*, and that sensation gave us a spring in the step which might last for hours. Each morning was the start of another adventure, every day an unfolding new horizon, every evening an unknown destination. Living outside, travelling the slow back roads, cooking over a stick fire, and drinking in the summer breeze and the starlit nights, was as naturally satisfying as anything we had known.

Charlie the pony was willing and fit, and we pushed him fairly hard, but we learned his limits, and in particular I learned to be very careful with the whip. When trying to get up some speed to cross over the A1 on a roundabout, we had to wait for about 10 minutes for a gap in the traffic, and in my hurry to get across before the next articulated monster truck thundered through, I used the driving whip to give Charlie a gee-up. I was a novice with the whip at that time, and the whiplash hit Ali so hard as it passed by her head that it nearly took her ear off. I doubt she has forgiven me even now, and to compound my misery the oncoming lorry driver saw what had happened and laughed like a drain "That'll teach you!" he yelled. He was right, it never happened again, but I still cursed him and his monstrous contraption, loud and long.

Another reason to be careful of the whip is that a good horse will work and work and work until it drops. A good horse would die for you if you ask it, and you might not need to lay on with the whip at all. If you need the whip, it is probably too late already. Sure, a whip is a handy thing for giving signals, or giving an occasional flick for a quick burst of speed, but a whip can do a lot more harm than good. I never did drive a horse to death, or anywhere near it, because I realised that you can make a good horse do anything you ask of it, and a willing horse relies on the driver not to ask too much.

Charlie would pull all day, so long as his feet were in good order and he had plenty of feed, water and regular rests. We would walk some of the time, although a steady trot is the preferred gait for horses when they want to cover distance, and on level going Charlie could eat up the miles. We learned the hard way the importance of grooming as we were heading home, when after a long day trotting across the plain of York, we pulled in to the gravelled car park at Fountains Abbey, where we planned to get the tea kettle on and take a break before pressing on to find a grass verge for the night. As soon as we pulled off the road, and before we had got out of the cart, Charlie went down on his knees, his back legs gave way and he started kicking and struggling on the ground. Alarmed, I jumped out and quickly cut his hame straps (horsemen always carry a pocket knife for exactly this purpose) which freed him from the traces so I could get him out of the shafts. Once freed, he calmly rolled over onto his back to scratch the itch which must have been tormenting him as he pulled in to the car park.

Governess Cart or 'Tub Trap'

Most work horses appreciate good grooming, which not only relieves the itches, but reaches the places they cannot manage themselves. Even the giants, the 18 hand Shire horses, will lay down and roll on their backs when let into a field after work. If there is no field available, after a day's hard pulling a horse will appreciate a few minutes spent scratching and massaging the pressure points where the body takes the rub of the harness - shoulders, saddle and girth, britching and the top of the head. Many horses go into a trance state when you do this, and good grooming avoids having to sort out the muddle if they go down on the ground and try to roll when still yoked up in harness.

We learned the hard way about mending the inevitable broken wheels and axles on the roadside. On our way North, we were trotting happily down the Clitheroe bypass when we were amazed to see an iron hoop rolling down the road ahead of us. It was our own iron tyre, which had become detached from the dried-out wooden wheel but remained upright and was now rolling away from us, going faster than we were. The wooden wheel stayed in one piece, just about, but we were riding on the wooden felloes. I had to gee-up Charlie to catch up, leap out of the back while Ali drove on, and then sprint ahead after the tyre to catch it. We fixed it back in place temporarily with nails and wedges, and travelled a few miles to Sawley where we found a stopping place. We jacked up the axle on the big tin trunk, took off the wheel, and soaked it overnight in a river so as to swell the wood enough to grip the tyre.

We travelled a couple of hundred miles on that tyre, soaking it frequently, doing 30 miles day, and using nothing but grass for fuel. We learned always to carry a few bits of metal plate, pre-cut, bent, wrought or drilled to various special shapes, to get us out of the inevitable troubles. Drilled plates of mild steel, 6" x 2" x 1/8th" would repair a shaft or maybe an axle. Chunky iron wedges two inches long would hold an axle box in place, and a set of slender three-inch iron wedges, bent over at one end with a hammer, was what we had used to fix the tyre. A pair of these could be hammered in between a loose tyre and the oak felloe, and then bent over on the inside of the wheel to hold it all together.

Apart from loose tyres we had to contend with loose axle-boxes, loose horseshoes, rubbed shoulders, hostile policemen, inconsiderate traffic, pouring rain, wet firewood, broken harness and painful blisters. On the other hand we were on the road again, and we drank in the slow panorama of the changing landscape, the smell of horse, the sights and scents of the hedgerow blossom and the breeze on the hay fields, the quiet weariness of laying on the bare earth beside the campfire at night, the sparkling joy of the dawn chorus, the priceless camaraderie and the long sweet song and steady rocking rhythm of the open road. After the exhilaration of our trip to the Yorkshire Moors, we were hungry for more.

We did not find our mysterious forest horseman in Dalby Forest - we simply ran out of time meandering along the back roads, and reached the point where we had to turn for home before we found him, but we had taken another big step - our mission was confirmed, and we were hooked. We knew what kind of horse we wanted, and even though we had little money to buy one, no grass to feed one, and no horse work to go to, we were full of optimism. Our project seemed impossible, but it never crossed our mind that our trip was wasted - it was another trial run, a deeper dive into the pool of the unknown. When we emerged, exhilarated, we felt ready for anything, and dived straight back in.

Walter feeding horses

5. Littlebeck - *getting started*

At Littlebeck (pronounced La'albeck in Westmorland) we became farm workers again. Two days after we moved in we cleared a wind-blown sycamore tree for our new landlord Gordon Jackson, carted it home in the tractor and link box, and filled the woodshed. Our house-mates Tony Ashford and Viv Lewis had a wee baby, Robin, but this did not stop them from getting stuck in to heavy manual labour. I have a faded picture of Viv unloading a trailer full of hay bales, with Robin strapped in a sling to her front, as if it was the most natural thing in the world, which of course it was, although Gordon was not so sure .

Within a week of moving in to Littlebeck and asking around, we were making progress. We found a job weeding young trees for a local landowner, Steele Addison and we toured around the forestry plantations nearby, talking to contractors to get an idea of who was who and what was what in the arcane world of forestry and timber contracting.

Within two weeks we had discovered the local traditional music scene, and we hardly looked back. One memorable night Ali and I set out for the music session in the pub in Morland, two up on one bike. We carefully crossed the ford over the River Lyvenet, taking off our shoes and rolling up our trousers. Once in the pub we had our first meeting with Colin Butterworth, a woodman from the nearby Winderwath estate. He shared our love of traditional Irish music and knew all the sessions and players for 50 miles around. Ali was a mostly a fiddler and I played the piccolo and tin whistle, so while we played the tunes he rattled out the beat on the *bodhran*, with a fine *tick-tickety-tick* style. Colin soon introduced us to singers, musicians, mighty sessions, ceilidh dances, and to woodmen, foresters and craftsmen of all kinds.

In December 1977, after a great night of music in Morland, ending with *Mull of Kintyre* playing over and over on the jukebox, and more pints of Marston's than was good for us, we set off two-up on the bike, back to Littlebeck. I was on the saddle, Ali on the carrier rack, and as we whizzed down the hill towards the river in the deep dark, with no lights and very little in the way of brakes, she yelled in my ear: "Don't forget the ford!" I did not even have time to say 'what ford?' before we hit the water at speed. We managed to stay upright, and laughed all the way home.

At Littlebeck with Tony Ashford and Viv Lewis.
Bill and Ali Lloyd, and Tom Lloyd, (and the beehive.)

At Littlebeck we earned our rent by occasional farm labouring but we focussed on our main mission - prospecting for forestry work. After a tip from a local contractor, one day we knocked on the door of the Forest Office of Pulford Forestry on Whinfell, near Temple Sowerby. We explained to the manager, David Lee, that we were looking for tree work, and in particular we were looking for horse extraction work. By one of the great coincidences which punctuate this story, we were told that Pulford Forestry had an operation in Wales, where they had an experienced timber horse. The horse was about to go to the knacker because there was no-one to work him. Would we like to buy him? (Do fish swim?)

We found out all about him – around 14 year old, a little over 16 hands, a Suffolk/Clydesdale cross gelding. Name of Ginger. Quiet, very experienced, ready for work. I did a quick calculation as to his likely weight, and I offered 'knacker price' for him – £400.00, unseen. (Knacker price is based on the market price of horsemeat, usually for export to Europe, where it is eaten like beef.) We knew that a good working Clydesdale gelding would make £1200 - £1400, but we did not have that kind of money. In fact we had hardly any money at all except for a tiny balance in a current account, but we also knew that Pulford had no further use for him and were ready to send him to the knacker. To our wonder and amazement, they accepted our offer. Pulford had an extraction job for him in the Eden Valley, but had no-one to work him, so the deal was that we would buy him, they would transport him north, and we would work him on their contracts. We shook hands on the deal, and left Whinfell in great excitement with a promise that the horse would be delivered 'within a few weeks.' There was a moment of doubt when we wondered if it was madness to buy an unseen horse, but only a moment. We had gambled and rolled a double six.

There is a quotation from W.H. Murray, quoting Goethe, which I have included at the start of this book. It concerns the moment of commitment, which somehow seems to start an inevitable process in which apparently unconnected events start to move towards the same point independently, and then meet each other in a benign synthesis which was unforeseen. We usually call that synthesis a lucky coincidence, and our path into logging with horses seemed to be a succession of such moments. I became interested in why this should be so, and would happily discuss it with anyone at any time, but I never found an answer, or even a theory, that explained it. Academics might call it synchronicity, Buddhists might call it karma and explain it with reference to re-incarnation; mathematicians, and astrophysicists who propose parallel universes are surprised that it does not happen more often, and philosophers debate as to whether one event causes the other, or whether there is such a thing as fate or destiny, or whether it is after all, mere coincidence.

Did the moment of my commitment to becoming a horseman influence the moment which decided that Ginger was to be sent for slaughter, which in turn meant that I could afford to buy him? Impossible. But that moment certainly influenced the fact that I knocked on the door at Whinfell looking for a timber horse, which in turn decided that he would not be sent to the knacker after all. Those events were connected by an idea, and one event was clearly dependent on another, by simple cause and effect. If I had not turned up looking for a horse, Ginger would probably not have survived. Common sense dictates that it was simply good luck, but whatever it was, those moments became inextricably bound together and influenced both our lives.

We had committed to our path when we quit our jobs and took to the road in tub-traps and bow-tops, and Providence began to unwind to help us along. Within a few months we had bought Ginger, and more doors opened for us. It is possible to trace that single clear line of cause and effect in a series of events, started by the moment that Walter bought a loaf of bread from Tony Ashford at Trentishoe Fair several years before, but that does not explain the coincidence of the two separate strands coming together with no apparent connection, and yet so vital to our enterprise. After a while I stopped trying to find a rational explanation, and was content with the knowledge that it really did happen. Goethe's description of the process is more pleasing and more optimistic than mere luck.

Meanwhile - now, suddenly, at last - we had a horse, although we had not seen him yet, and everything had to be done at once. We needed to find £400.00 to seal the deal, as we were anxious that David Lee at Pulford might change his mind, or get a better offer, or lose interest and send him to the knacker anyway. Back at Littlebeck we told Tony and Viv our news, and in another unlooked for, generous, and essential development, Tony immediately offered to lend us the money. We accepted, went straight to the bank, back to Whinfell, and laid our money down. We stepped over the threshold and into another world. We joined the 'people of the horse'.

Once Ginger was paid for we were keen to start, so we went with David Lee to inspect our first job - a clear fell of Scots pine and Sitka spruce in St. Croix Wood, on the ridge tops above Maulds Meaburn, between Orton and Appleby. The job had been started by Pulford's own tractor man, but the ground was too wet, or maybe his tractor had broken down, or maybe they could not agree a price, but anyway he had quit. The job looked feasible - the terrain was level, the price was so-much-per-ton piecework, and there was a grassy area in the wood where Ginger could graze. The only problem with winter coming on was lack of shelter.

We found from somewhere a large sectional shed to use as a stable, and borrowed a tractor and trailer to get it moved up to St Croix. Ali remembers well the day we erected it. We managed to get the walls up and bolted together, then I climbed up onto the gable and shouted down to Ali: 'Right, pass the roof up.' Tears and recriminations followed, on both sides, but we got it up in the end and we, (and the shed) managed to stay together, just about. Constructing a large wooden shed on a hilltop in the middle of winter with hardly any tools and even less experience was the most demanding job we had yet attempted, but even that soon faded in importance as we moved on to serious work.

Acquiring a chainsaw was the next mission, and once again we found that doors opened, people were generous and willing to share their knowledge. In King's Meaburn was a forestry contractor who worked in Greystoke Forest, and he sorted us out with a good old Husqvarna chainsaw. He showed me how to start it, how to sharpen it and maintain it, and how to tell a good one from a bad one, (compression). He showed me how to get the fuel mixture right - ("never mind the manual, just add 2-stroke oil into the mixture until the exhaust gets so smoky that breathing it in becomes a problem.") He worked on a mixture of 20:1, telling us that the saw would last twice as long. He was probably right – I still have that saw 40 years on. He fitted the saw with a guide-bar about 3 inches shorter than the recommended length, telling us that a short bar was more efficient than a long bar – more power per tooth, and less likely to cut your feet off when snedding. He would keep a separate saw with a longer bar for felling big stuff. The short bar gets very hot, needs regular greasing, does not last as long, and is strictly contrary to manufacturer's warnings, so I would not recommend it, but I still have all my toes. We had to borrow more money to get the saw, but we got it, and I spent the next few days stripping it down, putting it back together, starting it up and sawing firewood for Gordon Jackson. We took a trip to Rickerby's in Penrith to buy safety helmet and gloves and special non-fling oil for the bar, all of which were frighteningly expensive, but I was ready for work.

The fact that I had no training did not bother me at all, although it should have done. There are now (2020) good training courses for all aspects of forestry work, at reasonable prices, and a chainsaw operator's 'ticket' is an essential passport to employment. Although I learned on the job by watching experienced operators, that involved a certain amount of '*macho bravado*' which I would not recommend. With hindsight I was lucky to come out of the wood in one piece, and had a few brushes with serious injury, so my experiences should certainly not be used as a model for aspiring woodmen, but that was how we did it.

Now we needed wheels. All we had were bikes, but working horses need to be fed at least once a day even when not working, and at least twice a day when pulling wood. We now had to carry feed, harness, and a chainsaw. St. Croix Wood was a few miles away, and although we cycled there and back for the first few days, the winter weather was approaching and we could not keep it up. We could not monopolise the shared Morris 1000 van, which belonged to Tony and Viv, so we asked Pulford if we could put a caravan at St. Croix and live in the wood. They were not keen, but instead they found us some wheels – an old Renault 4 van. In return for the van, and while waiting for Ginger to arrive, we agreed to do a scrub clearance job at Cowan Head, by Ings, near Burneside, about 20 miles away. Cowan Head wood had been planted with Sitka spruce, but the gorse (or *whins* as gorse is known in Westmorland) was encroaching faster than the trees were growing and was smothering the young plants. Our job was to cut back the whins using a brush-cutter and chainsaw, stack them in heaps, and then burn them. Back at Littlebeck that evening we gathered up the brush-cutting machines, chainsaws, fuel and sharpening tackle, the much prized waxed-cotton waterproofs, (known as Solway Zippers,) packed lunches, pitchforks and waste oil for the fires, loaded up the van, and set the alarm-clock for an hour before daylight for our first contract work.

With two flasks of coffee, wrapped in woolly hats and mittens, and freshly dubbined boots, we scraped the ice off the windscreen of the van, warmed up the engine, and set out through the frozen Eden valley landscape. The snow was crisp, and the moon and stars were still bright in a cloudless winter sky. The excitement of having our own van was dampened when we reached the first hill – the van was overloaded and underpowered and could not get up the steep and slippery slope. After several failed attempts, I recalled reading somewhere that reverse gear was usually lower than first gear, so we turned around and tried speeding up the hill in reverse. It worked! For a few days we had the high hilarity of leaving home, reaching the steep hill, making a quick three-point turn, reversing up at speed, then turning round again and pressing on. We kept that up until we found that there was in fact a (hidden) first gear!

Away we went, off through the icy wastes, up over the A6 at 1,400 feet past Shap Summit, for two weeks of hard graft cutting and burning whins in the ice and snow. Most days were cold and foggy, and we worked in a strange atmosphere of mist, two-stroke exhaust fumes, the resinous smell of the cut whins and smoke from the bonfire, all swirling among the rocky outcrops and the dark young Sitka trees. I used a brush-cutter and chainsaw to cut the stems, Ali used a pitchfork to heap them up, and we ate our sandwiches sitting in the snow round the bonfires. Our wages paid the rent and repaid part of what we had borrowed to buy Ginger, Pulford were happy with our work, and we were impatient to get started on the big stuff. All we needed was that Welsh horse.

Ginger, 1978

6. *Ginger* - *one horse power*

Soon enough we had the call – Ginger had arrived from Wales and was in a field at Whinfell, near Penrith. We shot off like sprinters from the blocks, and on a bright crisp morning we parked outside the field gate. It was a 30-acre field, surrounded by Sitka spruce and at first we could not see him and thought he had escaped already – disaster! – but then we spotted him away in the far corner. We called to him, climbed over the gate, rattling our bucket of feed, and he set off towards us. We met in the middle of the field, in a moment of bliss – realisation of imagined glory. He was a cracker. Big, powerful, fit, alert, very friendly, totally biddable, bombproof, gorgeously coloured, hard feet, good steady gait, good teeth, and with a big hairy bottom lip that I would come to love …

All we had with us was a rope halter, but we slipped it over his head, and led him around for a while to get to know him, but we could do nothing without harness. The working gears that came from Wales with Ginger were barely usable – a broken collar, a piece of torn canvas for a back band, and two unmatched trace chains. We had to find a good set of harness before we could start. Decent gear was hard to find and so was expensive. First we had to decide which type of gear to use.

I was familiar with the trace-horse harness that I had used at the Morton's. The traces are the ropes, chains, wires or leather straps that do all the heavy work in a set of harness. One end of each trace is attached to the hames on each side of the collar, and the other end attached to the load. Trace-horse gear is used when working two horses in tandem to pull a vehicle with shafts, and the traces of the front horse (the trace-horse) are hooked on to rings at the front end of the shafts. The traces on the shaft horse hook on to metal fittings at the other end of the shafts. The shaft horse steers the vehicle and the trace-horse provides extra traction. (Before motor lorries replaced the working horse it was normal practice for a farmer who lived at the bottom of a steep hill on a main highway to keep a spare trace-horse available, to help pull heavy loads up the hill, in return for a small payment.) Trace-horse gear comprises collar, hames and traces, a back band and a crupper, with extra straps over the hindquarters to support a high spreader bar. The spreader would hang in the gap below the buttock and above the gaskin and hock. (A bellyband is optional, but I did not usually use one in the wood as it tended to catch on the brash.) The chain traces extend back for about a yard from the spreader bar, each with a hook at the end to attach the load. (See Appendix 2).

The alternative to trace-gear is 'snigging gear' – a simple back band with chain traces, dragging a swingletree with a single central hook. The full trace-horse gear has some advantages – it can give extra lift compared to the more usual dragged snigging gear, so the front of the load can be lifted slightly off the ground. Also the spreader itself does not catch on the ground or on the stumps like a swingletree. The disadvantage of trace-horse gear is that the upward lift puts more pressure on the horse's back legs, and there are more straps to break. On-the-job repairs to harness turned out to be one of the biggest causes of 'downtime' and because most of my forestry contracting in the early days was done on piecework, not on hourly rates, time was precious and breakages were expensive. The fewer straps, the fewer breakages, and the less downtime.

The horse's collar must fit exactly, because otherwise it will rub the shoulders and blister the skin. A blister may not be evident at first, but it will be painful, and the horse won't want to pull, so a work horse must be checked carefully each day. It becomes a habit while brushing to run the hands all over the sensitive areas of the shoulders before yoking up, looking for any swellings, lumps and bumps. The shoulders, belly and back are the usual places where rubbing occurs, but the less visible areas should also be checked - under the tail where the crupper fits and under the top of the bridle between the ears can both get rubbed raw if the gear does not fit. If a blister bursts it is not pretty, but more importantly it is very tender and liable to infection. Salt water is the best first-aid. A rub with methylated spirits is said to harden the skin on the shoulders, and I did use it occasionally on the road, but Ginger never had sore shoulders in the wood, so I guess he was hardened to the work before he came to me.

Ginger in trace gears, 1977 (See Appendix for names of parts)

Most of my working horse gear was found at country auctions, or at farm sales, where it had usually been hanging up in farm buildings for the previous 30 years. The leather would be dry and the stitching rotten, but with care and the liberal use of harness oil it could be softened up and brought back into use. Flexalan and Hydrophane oils were the best, but expensive. Neatsfoot oil was cheaper, but there was disagreement among horse 'experts' as to whether or not neatsfoot oil rotted the stitching. That was not so relevant to me because most of my gear was old and needed re-stitching anyway, so I preferred the cheaper neatsfoot oil. I soon learned how to stitch harness - a heavy duty leather punch, a selection of awls, a packing needle, a roll of hemp thread and a block of beeswax were the essentials, with plenty of time and even more patience - neither of which I had in abundance. But since I was paid by 'piecework', time and patience had to be found, otherwise no wages, so I spent hours with awls and needles, beeswax, hemp and broken strapping.

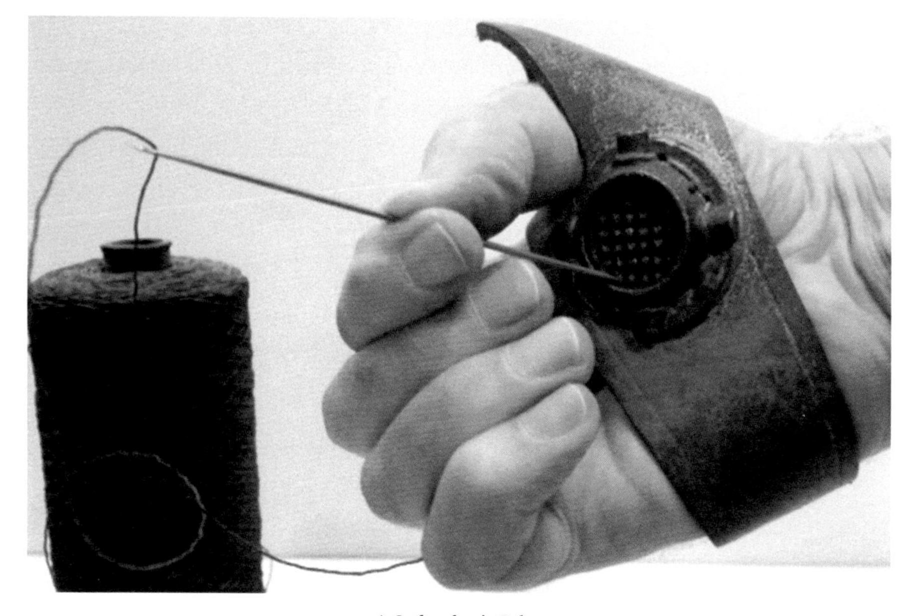

A Sailmaker's Palm

I used a sail-maker's palm save injury to my hands for the tougher sewing jobs like heavy leather traces. A 'palm' is a flat thimble, a circular plate fitted in a leather strap, worn like a glove. It allows the whole strength of the arm to be applied to the needle as it passes through thick leather. There are various useful proprietary gadgets for double stitching, where two threads are passed in opposite directions through the same hole. I acquired a *Junker and Ruh* leather sewing machine with a 12" lever, designed for stitching leather soles onto shoes, and it could sew anything. Eventually I wore it out, but it still sits in my shed waiting for a new bottom bobbin and some curved needles. (Leather machine needles have a channel from the eye to the tip of the needle, to carry the thread, which is linen or hemp, pointed and stiffened with beeswax.)

We put the word out that we were looking for harness, and soon we found what we needed. Edward Holliday who farmed near the A6 south of Penrith sold us a set of trace gears that he told us had come from Lowther Castle, and a collar that had been his father's, all for £15.00. Colin Butterworth sent a message to say that he had heard of someone who lived near Kendal who had a barn full of gear. After some weeks we worked out that it must be Edward Acland of Sprint Mill, who kept two tool shops in Kendal – 'Tools Past' and 'Multum in Parvo' – both of which were piled high with the pick of his extraordinary collection of useful bygones. My diary note at the time is unusually full of detail about our first visit to Sprint Mill, and the richness of his Aladdin's cave. We did a deal for a collar and a few straps, and we became firm friends, remaining so to this day.

An alternative to revitalising old leather harness was to use seat-belt webbing, and later I had a set made up by Bill Moore (Lifting Tackle) Ltd. in Heysham. It was made to measure, almost unbreakable, lightweight, easy to clean, and easy to stitch. It was not cheap, but still a quarter the price of leather harness. (A DIY set made from scrap-yard seat-belts would cost next to nothing.) I preferred the look of my leather trace gears, with shiny brass buckles and decorative plates, but seatbelt webbing was a very practical and inexpensive alternative. (See Appendix 2 for a diagram *Parts of The Harness*)

Ginger needed a 27" collar, and since a new one would cost many hundreds of pounds and so was out of our reach, I only ever used old ones. Sooner or later they fell apart, usually after a serious jam which broke a hame strap and ripped the collar as the hames flew off. I would keep at least one spare collar, but was always on the lookout for replacements. Once or twice we would have a good collar relined – at a cost of around £80 – but more often I would pick up a 'new' old one for £30-£40.

Once we had the harness, we needed chains. Finding the right weight of snigging chain for the load was a challenge. Too heavy a chain was clumsy, awkward to handle, and added a significant weight, but too light a chain would break and tangle too easily. There was no exact specification, but soon enough I would know just by handling it whether a chain was suitable or not. Different horsemen have their own system for yoking up the horse to the load. The system I learned from Joss Rawlings was to use a snigging chain about eight feet long, with a hook on one end, and a heavy ring about 4-5 inches in diameter on the other end.

Like most specialist work, forestry has its own jargon. Woodmen seldom call a tree a tree. A standing tree is called a stem, or maybe a butt. When a tree has been felled it is called a pole, or with typical understatement it is called a stick. (For real effect, a woodman might call a 200-foot Douglas Fir measuring 6 feet across at the base and weighing a few tons 'a decent stick'.) When dragging timber, the wide base is known as the butt and the top end is called the tip. If the load is a single pole, it can be dragged out either butt-first or tip-first, and different terrain would dictate which was the better choice. 'Tip-first' produced fewer jams, although the flexibility of the pole could make it harder to extract round standing trees, because the tip might break off. 'Butt-first' allowed the traction to be applied to the heaviest part of the pole, which was more efficient, but a jam was more likely and more serious. More often than not the choice was dictated by the woodcutters, who would fell a tree in whatever direction they could get it down most easily, and the horseman had no choice in the matter. When I was cutting my own wood I would decide at the start how the trees should go and then try to cut them all the same way, and drag them out the same way, which allowed me to build a proper stack at the roadside.

If the load was to be multiple poles, unless they were very small first thinnings, they would normally come out tip first, pulled into a loose bunch by hand so that the tips were all within a few feet of each other. The load chain would then be wrapped round each in turn, and when the horse began to pull they would be dragged into a tight bunch. Making up the load in this way could be strenuous work for the horseman, but

First Thinning, using snigging gear.
(Note the temporary repair to Ginger's nose band. He deserved better.)

there was little alternative. The woodmen had an old saying 'the smaller the wood, the harder the work', which applied to this job more than any - with small poles it was a waste of time to yoke the horse to a single pole that could be moved by hand. Since it might take a dozen small poles to make up a load, they would be bunched up by dragging them by hand, which is much heavier work than felling or stacking. Timber tongs and strong gloves gave some necessary protection, as more of my minor injuries were caused when gathering the load than by any other job.

For larger poles, too big to move by hand but too small to be worth extracting singly, the technique is different. The snigging chain is attached to the furthest pole which would be dragged forward until it is level with the next pole. The chain is put round the next one with a half-hitch, then both are dragged to the next, pulling tight each time, until there is a worthwhile load and the chain linking them all together is as short as possible. For a load of many poles, this might use the whole of the chain, but usually it would leave a few feet of spare chain between the ring and the load. The ring itself is then used to take up the slack and shorten the chain so as to get the maximum lift on the butts and the minimum length on the chain, using a clever loop that can always be undone (hard to describe but easy to demonstrate). Finally the ring is hooked on to the trace hooks, and all is ready to pull.

A variation I used on snigging gears (not trace harness) was to have a slot cut into a flat hook in the centre of the swingletree, instead of using the ring on the chain. The horse was backed up to the load, while the horseman carried the swingletree, and the snigging chain was slotted in to the hook at the right length. I used both systems frequently, but because trace harness was more liable to breakage I tended to use a swingletree in difficult going, and would always keep it as a spare even in good going.

I spent some time in the library at Newton Rigg Farm School researching alternative methods of horse extraction – the Norwegian skidding arch and the Canadian sledge bolster. The skidding arch had a clever quick release harness which attached the end of the shafts to the breast collar with two pins, one on either side. The purpose of the arch was to lift the butt off the ground, thereby lessening the friction and avoiding jams. The Forestry Commission had done some careful research into horse skidding arches, and as a result they imported a number of them for general use. I called their research station at Alice Holt Lodge, and discovered that there was a complete skidding arch at Grizedale Forest, lying unused in a shed, so I called the Grizedale Forest Office. They confirmed that it existed, but they were not sure if it was for sale. We made a mental note to follow it up next time we were in the area.

The purpose of the sledge, like the skidding arch, was to lift the butt off the ground, lower the friction and avoid jamming on stumps, but the sledge had a lower centre of gravity than the arch, so was more stable. The sledge looked like the better bet for me, particularly because I could make my own, combining a Forestry Commission design with one from a comprehensive Canadian book, *Pulpwood Hauling with Horse and Sleigh. (1943).* The principle of the sledge construction was to make all the wooden joints loosely, but with iron bearing surfaces, and with enough play in the bolts so that each of the sledge runners would move independently up and down and side to side. Oak heartwood was needed for strength to make the bearers, which carried the weight, and poplar was used for the sledge runners because it did not split or tear easily. I found a maker of fine furniture nearby, Ian Laval, who had a stack of seasoned wood, and he supplied what I needed at a generous price. The blacksmith in Morland made the steel plates for the joints and the runners, and after a couple of satisfying days sawing, drilling and hammering in Gordon Jackson's workshop at Littlebeck, I had a bolster sledge. Ready to roll.

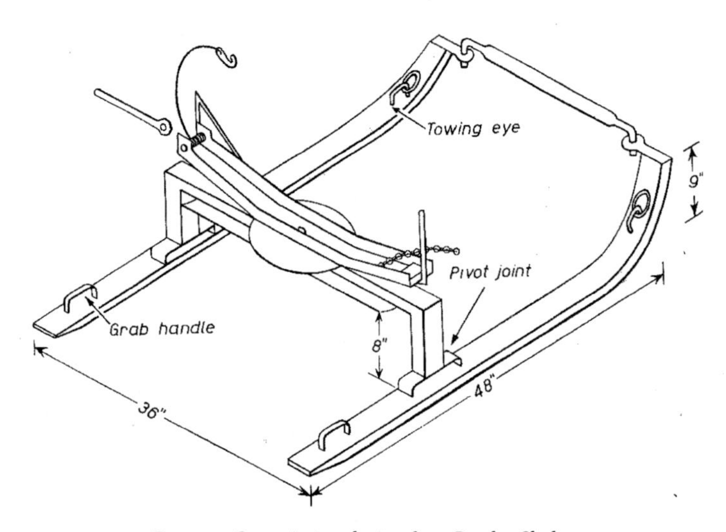

Forestry Commission design for a Border Sledge

7. St. Croix Wood - *an awakening*

So we started. I walked Ginger the 10 miles from Whinfell to St. Croix Wood, and that walk remains a vivid memory, an imprint triggered by the extraordinary exhilaration of passing through the vast Eden Valley landscape on a fine bright morning, with a fine great chestnut horse, on my way to my first horseman's contract, trading on my own account. After a year of planning, the reality and the dream were one. In those first days, the simplest of things were full of pleasure - coltsfoot and catkins, a fox in the wood, an old tweed jacket from a charity shop, a Belstaff waterproof, battered moleskin trousers, a sleeveless denim waistcoat, and of course a wide brimmed hat or a woolly bobble-hat for winter and a good pair of boots. (I tried wearing 'tackety boots', with soles and uppers rigidly stiff and around 100 hob-nails in each boot. Some Westmorland shepherds never wore anything else, but they never suited me.) Most important was an ex-Air Force gas-mask bag holding home made bread, home made cheese, home made fruit-cake, and a flask of coffee. I dressed like a young fogey, but did not ever think of it as a fashion statement - these were the right clothes for the job, and I didn't care if I looked 20 years older! My charity-shop clothes were the marks of my new trade, and of our status as hippy activists, worn with pride.

After a few hours walking, Ginger was turned loose into the wood at St. Croix to graze, and I surveyed the trees. They were alarmingly tall - much bigger than anything I had felled before, but there was no-one watching so I could practise my chainsaw technique privately, and after the first few trees I could drop them to within a yard of where I wanted. Not perfect, but good enough. Tree felling itself is mildly technical, but once the technique is understood it is not too strenuous, especially with a good sharp saw. The dressing out, or snedding, however (taking off the side branches after the tree has been dropped to the ground), is sustained hard graft - the most tiring part of the job and the most dangerous, so good technique is essential. The chainsaw bar is often out of sight, hidden beneath the branches, or held at an awkward angle in order to get at the branches underneath the trunk and cut them cleanly. The woodcutter's feet are also out of sight, and usually within just a few inches of the end of the saw. All I had for technical guidance was an invaluable series of Forestry Commission safety leaflets, which set out the safe sequence of felling (preparation, cutting the gob and making the back-cut etc) and snedding. (Mind your feet, and mind the tension in the wood.)

The chainsaw was fast and efficient, but I disliked the noise, the fumes, and particularly the vibration. After a while I tried using the Yorkshire bill-hook for dressing out wherever possible - it had a long handle, and enough weight to get through most side branches with a good swing. The bill-hook used no fuel, produced no smoke, and needed less frequent sharpening, so sometimes it was more cost-efficient to dress out by hand than with a chainsaw. On first or second thinnings where the poles grew tall and straight with fewer and thinner side branches, some species could be dressed out by knocking off the branches rather than cutting them off. Young larch is particularly easy to dress and can often be extracted without dressing out at all, as the friction along the ground takes off most of the underneath branches and the remainder can be taken off with a bill-hook in the loading bay. Scots pine was also good to handle - few side branches, soft wood so easy to cut, and the soft needles are easy on the hands. St. Croix was my first experience of Sitka spruce, and I hated its prickly brash. I commented to my boss at Pulford how I disliked it. "You get used to it" he said. "I don't want to get used to it!" "Tough," he said. I got used to it, helped by a good pair of gloves.

After each tree had been dressed out, the brash was piled up carefully in rows so that all the tree stumps were visible. When a few poles were down and laid out in the right direction for extraction, it was time to yoke up. This was the big moment - I had worked horses before, pulling carts and wagons on the road to Appleby, or in the fields in Yorkshire, but this was the business, with my own horse on my own contract. This was when I would find out how good Ginger was, how good I was, how much weight we could pull in a day, and if that was enough to make our living.

With Ginger at St. Croix Wood, above Appleby.
(Note canvas back band, with sledge bolster just visible.)

On the first day we felled, dressed out and extracted a dozen trees, converted them to specification and stacked them at the roadside. At first we worked with long reins, (known as cords) but before long it was obvious that Ginger knew what he had to do and so we could work by voice commands. I had to learn the right vocabulary - Yorkshire horses and Welsh horses are taught different words for left and right!

In Yorkshire they said *'Arve Up'*, and *'Gee Back,'* which Ginger simply ignored, but he did respond to *'Get in'* and *'Get Back'*. The words for stop and go were the same – *'Whoa'* and *'Go on,'* or better still *'Get on'*. I found that tone and inflection were almost as important as the words used, and before many days had passed he would do as he was asked. Ginger was clearly more experienced than I was so I watched how he worked, went along with it and probably learned as much from him as he did from me.

The previous contractors had left high tree stumps, or stocks as they were called by the local woodcutters, and these high stocks were a constant problem. If the stocks were not cut low enough to the ground when felling they usually got in the way when extracting the poles. The load would jam up against a high stock at speed and something would break. Usually it was the hame strap, so that strap was important – the hame straps had to be strong enough to take the strain but if the load jammed against a high stock the strap had to be weak enough to break before the hook, before the chain, and before the horse's shoulder.

To avoid the stocks and the jams we extracted the first few loads using the new sledge, but some of the butts were just too heavy for me to lift onto the bolster. To keep the butts from jamming without using the sledge I tried using trace-horse gear, which lifted the front of the load as Ginger pulled, but that gear put too much strain on his back legs. I changed to simple snigging gear, comprising back band and swingletree and also changed from dragging butt-first to dragging tip-first. We had fewer jams and Ginger was immediately more comfortable and relaxed and with a combination of dragging tip-first and cutting low stocks, production increased. At the end of the first day there was a respectable number of poles in the loading bay, ready for conversion into saleable products. By the end of the week there was 10 tons - half a wagon load.

For the whole of my time in the woods, there was a constant difficulty in finding woodcutters who understood that felling trees for horses is a different job from felling trees for tractors. The chainsaw man always wants to cut the stocks as high as possible, because two feet above the ground the tree girth is less, so by cutting high stocks there is less diameter of wood to cut through, less expensive fuel is needed to run the saw, and less downtime for refuelling. High stocks also mean less grit in the chain, and so less time lost in sharpening. The horseman on the other hand wants the stocks as low as possible, because jams happen easily even when pulling the poles tip first, and a jam often means a breakage. Even if nothing breaks, a jam is always hard on the horse.

St. Croix Wood, Maulds Meaburn. Carrack Fell in the distance. (Photographer unknown)

Apart from wanting low stocks, the horseman wants the minimum of obstacles in the form of brash and tops. When cutting for a horse, the chainsaw man must take the time to clear the brash out of the way before moving on to the next tree. A chainsaw man working for a tractor will want to fell the trees as fast as he can, whichever way they will fall easily, and to leave the brash where it lies, knowing that the tractor and winch can work through a mound of brash, hook on to a butt and drag it clear. Muscle and bone must work more subtly, so the horseman is frustrated by the extra time taken to clear a jumble of trees in a jungle of brash.

If the woodcutter and the horseman were both on piecework rates paid by the ton, these were incompatible issues. In a bad jam on a high stock, the horseman might break one or more straps or shackles, which might take a half an hour to fix, even assuming that he carried spares in the toolbox. (I learned to use lightweight shackles as 'weak links' like the hame straps, so that they would break easily in a jam, like a fuse in an electrical circuit.) On the other hand, the chainsaw man might take an extra minute or two to make the felling cut at ground level because of the extra girth, or waste 10 minutes re-sharpening due to grit in the saw, so cutting low stocks could cost him half an hour a day. Half an hour a day for either man was a significant loss, so a balance had to be struck. That balance could be hard to find, especially if there was more than one woodcutter in the gang, as they blamed each other if I complained that the job wasn't tidy. I soon realised that my margin of profit was so small that I could only operate efficiently if the woodcutters worked for me, and not the other way round. Because good woodcutters were scarce I would usually fell my own wood, until I met Dallas Machell - the best man I knew who liked to cut wood for horses, and who was willing to take the time to get it right. More of Dallas later.

It is difficult to explain to anyone who has not tried it why jams and snags happen so often. The forces, tensions and pressures at work when a horse drags a load of poles out of a wood over difficult terrain can be complex and unpredictable. Most of the film of horse extraction does not show the inevitable jams and snags, but a critical part of the horseman's skill is to avoid them or else to overcome them. Horse extraction is as much an art as a science and it was pride in the art, the craft skill, that made the job so satisfying. The obvious snags like high stocks, sharp bends and low branches can be identified, and the horseman can learn to judge whether a pole is too long to get round a curve or too wide to get through a gap. Man and horse must also develop an 'eye,' an instinctive quick judgement, part logic and part intuition, about where, in four-dimensional space-time, the two ends of a pole will be at any given moment. This 'eye', this intuitive feel for how the load will behave, makes a difference when pay day comes - and a good eye can keep the horseman out of hospital. Sure, a tractor man needs good judgement, and for all the same reasons, but many of the subtle judgements can be made irrelevant by some extra hydraulics, a bigger tractor, and longer winch ropes. Learning the subtle techniques of the horseman while working up-close with the living, breathing forces of nature, rather than battering them into submission with a machine, brought me the beginnings of a new understanding and a new way of looking at the world.

On my first job at St. Croix, I experienced something of the mythic nature of logging with horses, something subjective, possibly romantic and idealized, but grounded in the material world. Ideologically I aspired to be a part of a natural, self-sustaining ecological cycle, and I became fascinated by the relationship between the ideology, the craft and the science; between philosophy, intuition and logic, and with the basic physics at the heart of it all. Horse logging also became a political matter, concerning the ownership of the land and the trees, the relationships between the owner and the contractor, between capital and labour, and the nature of the power needed to harvest a crop of timber.

At St. Croix I began to see the woodland not as a commodity or a production line but as an 'otherworld', another plane, like another planet where I had to learn new rules and new ways of being. Working in the wood convinced me of the impossibility of 'owning' natural processes, and made me realise that 'ownership' is a legal fiction – accepted as true but in reality impossible. I developed a sub-conscious understanding about the relationship between living matter (in the form of timber), energy (in the form of horse traction and sunlight) and space (in the form of the woodland). I could not express it as a mathematical formula, my language was inadequate to describe it exactly, but I recognised the inability of human consciousness to grasp the infinite complexity of the natural world. I discovered that some of the essential facts of life - the nature of growth and the miracle of the harvest - lie beyond the reach of words, accountancy and economics, in a different, miraculous, dimension.

After the Ice Ages, before any trees appeared, the space above the ground at St. Croix had been just that – empty space. Trees seeded themselves or were planted, trees grew, time passed, and the empty space became solid matter in the form of 80 cubic metres per hectare of heavy timber. The highly evolved and sophisticated chlorophyll molecule used the photon energy of sunlight to provide food for self-replicating cells of DNA by making oxygen and glucose from carbon dioxide and water.

Photosynthesis: a leaf of *Bryum capillare*, which converts carbon dioxide and light into matter.

That seemed to me like some kind of miracle for a start. I was employed to harvest that miraculous stuff, pull it out of the chaotic wildness, where it had grown by a process beyond my comprehension. I converted it into a saleable product, stacked by the roadside to wait for a lorry to take it to a factory, where it would be processed into newsprint or chipboard, or maybe roof beams or toilet paper. The woodland became empty space again, the living mass was harvested and moved, one space was created and another space was filled. Even after the trees had been hauled out of the wood, the logs retained their miraculous, otherworldly character in my mind – a miracle of photosynthesis and evolution as much as a valuable commodity and feedstock for our industrial systems.

That insight into the natural world gave my work and my life a meaning beyond mere 'production'. The insight revealed part of what is really happening when we harvest the generous nature of woodland, and how sustainable harvesting practice is linked to the intuitive and instinctive relationships we have with the land. The 'ownership' of land gives legitimacy to the ownership of the harvest and the value of my stacks of timber could be easily stated as numbers - as cubic metres or tons or products or money in the bank, but the 'meaning' of the timber, its true and original nature, the essential reality of the log, was not made up of numbers, or products, or money values, or even of any words, except perhaps 'miraculous logness'. Words and numbers are mere symbols and the word 'log' is of course a word, not a log. The whole forest is 'other' - part of another realm beyond language and which for millennia has been understood in folklore and religion as a place of adventure, romance, danger, sanctuary and enchantment.

A log dragged out by a tractor becomes a commodity - plain industrial feedstock - as soon as it reaches the roadside. A log dragged out by a man and a horse - flesh and blood and bone - can remain a miracle, a wonder of nature, with a special meaning, even if only for the horseman. The true value of timber production for me was the miraculous, bountiful transformation from empty space into solid matter through time - a transformation which could be repeated over and over again, by photosynthesis and the energy of the sun. When I looked at a stack of wood with this insight I did not see a 'product' but an awesome manifestation of the same life-force and pure energy which created the horse and the horseman who then harvest the miraculous bounty. All the

Nature's way of fixing carbon.

pieces of the jigsaw were manifestations of the same energy, the same natural forces reproducing themselves over millions of years of slow evolutionary change. Tree, horse and horseman are part of one family, like distant cousins, all variants of the same extraordinary life-form that is the DNA molecule, about 35 percent of which humans share with trees, and about 80 percent of which we share with horses.

Guided by the signposts from Gary Snyder and Satish Kumar, St. Croix taught me a childlike wonder at life itself, and how precious is the life, and how precious the wonder. Using horses instead of tractors allowed me to preserve that wonder, that sense of fragility, of harmony, balance, and communion. Horsepower led me to believe that if we honour and respect that life-force and those natural processes, we may not be as productive, but our lives might be given new meaning, and we might at least endure as a species. Maybe, if we have a right relationship with the land, we might preserve our self-respect as well as our planet.

The logic behind this insight was unscientific and inadequate, and the economics were naive. The resulting description may be vague, sentimental and subjective, even romanticized, but I am not embarrassed or ashamed of my limited ability to express these thoughts, and the lack of empirical evidence and literary precision doesn't bother me at all now. My difficulty in expressing these ideas increased my respect for the great poets, who journey into this 'otherworld' of wonder and return, like pearl fishermen, with poems to prove that they have been there. My literary aspirations sprang from my association with the poets at the Arvon Foundation, but these insights arose from practical experience, in which the techniques of horse logging became a form of meditation or spiritual practice, accessible only while I was working in the wood. The insights would come to mind without any resolution or conclusion, while hauling logs, or contemplating a tree, or a landscape, or grooming a horse, or drinking tea from a tin mug, steaming in the cold air.

I knew that the discovery had changed me forever. I became more aware of my own literary and philosophical limitations, but to compensate I realised that meaning does not always require words. We can find meaning, harmony and satisfaction at a sub-conscious level, which we cannot always explain. My mission had taken me to another country, inaccessible to logic, in which the accountant's bottom line, the internal combustion engine and economic theory had no place and could probably never have a place. If I wanted to explore it, I had to tread lightly.

I also had to make a living, so had to use a chainsaw and a van and I had to keep my accounts up to date and show a profit, so of course my ideal world was a compromise with reality. I soon became aware of harsh economics, of the place of the labourer and his real relationship to the boss and the dependence of both boss and labourer on the accumulation of capital. My awareness of the political relationship between labour and capital did not make me any more or any less of a socialist, although I concluded that I was now, for better or for worse, probably a capitalist. I owned the means of production, which was the classic Marxist definition. I had borrowed money to buy the horse and the chainsaw and was now hiring them out with my labour, so I was trying to make a return on my investment as well as a living wage. Soon I was even buying and selling timber, so was exploiting the woodland for private gain. I accepted that made me a capitalist, but I wondered if a capitalist must inevitably be an exploiter. Could a capitalist not be a harmonious part of the natural world to which I felt that I belonged?

In order to maintain my self-respect, I needed to show that it was possible to be a capitalist without the worst characteristics – without being selfish, greedy, destructive, rapacious and brutal. Does it matter anyway? Does anybody care? Forty years later, when the credit crisis, the fiscal crisis, the banking crisis, the Euro crisis, the Brexit crisis, the global pandemic and looming environmental catastrophe all indicate that capitalism has some major defects, it seems to matter more than ever.

I did not find a satisfactory answers to the problems of capitalism, then or now, but I knew as soon as I started pulling wood in St. Croix that my political ideology and my spiritual home did not fit easily into any box, or with any label that I knew about. I had steered my own course against the rain and wind into unknown territory to try to change the world, but I soon began to doubt my old certainties and to retreat from ideology. My reading of Orwell and Solzhenitsyn had convinced me of the danger of totalitarian states, but I could not reconcile that with my conviction that unregulated markets and uninhibited consumerism had to be controlled. In the 40 years since I took refuge in the wood, the free market has brought enormous wealth for some, alongside devastating environmental damage. The increasing power of the state and social spending on health and welfare has lifted millions of the world's poorest out of poverty, and raised the living standards of billions more, yet capitalism remains both voracious and fragile. With hindsight, my expectation of imminent economic collapse 40 years ago was plain wrong, but as the planetary environmental crisis intensifies, so the need for a solution to the contradictions of consumerism becomes more urgent.

Most of my problems during those first days in the wood were less political and more immediate. After a series of breakages I found that antique trace hooks and ancient leather hame straps were not up to the job, so with my first wages I bought modern forestry hooks and new English leather straps with proper unworn buckles. As I worked out the techniques of handling the chainsaw and the horse, production increased and within a week or two we had settled into a pattern. We arrived in the wood early so as to feed Ginger an hour before he began work, and while he was chomping his breakfast I would sharpen up the saw, cut an extraction route, then fell trees for an hour before yoking up. After a light load to warm up, we would pull for an hour or two, then Ginger would rest while I converted and stacked the timber. At midday I would take half an hour for my own food, though there was little point in hanging around in the cold, especially when I was steaming and sweated up, so I preferred just to have a cup of coffee and a sandwich and press on. After the break we would do it all again for a second stint. In the icy weather we would ease off and pull smaller loads for the final hour each day, as it was not good to leave Ginger still steaming when the frost came down with the dusk, and we aimed to be out of the wood by 4 o'clock.

The terrain at St. Croix was mostly level going, so we could keep up a fairly steady pace. When we moved to the steep slopes of the central Lake District I discovered that the hardest part of extracting timber by horse (for the horseman anyway) was climbing back up the hill to bring the next load, and we would often need a longer rest at the top of the hill than we needed at the bottom after pulling it out. St. Croix was easy going and a good way to start, but it was soon done.

The timber wagon came once a week, loaded up our week's work in 10 minutes, and drove away leaving a stink of diesel exhaust fumes. Before the last stack was ready to go I was already wondering 'what next?' As the last wagon load disappeared down the road our job was finished and we came down with a bump. When we reviewed our position, our problems seemed to multiply the further into it we got. I now needed regular work to pay my off my horse debt, to pay the rent at Littlebeck, and buy the groceries. Somehow I had to pay for grazing, hay, hard feed and winter stabling for the magnificent Ginger. Pulford Forestry had no more work to offer, so we would shortly lose the indispensable works van, and the chances of finding a field for a heavy horse in mid-winter were very slim. Most hill farmers do not care for horses, which damage grassland, especially when the ground is wet, and this was Cumbria, where most livestock was taken off the land and into winter housing in November. Even in summer, grazing would not be easy to find - horses hang around in one place waiting to be fed, they dig out great clods with their feet when they gallop about for fun, and they drop their dung in the same place every day. Regular harrowing and careful management is needed for horse pasture, and most farmers have better things to do.

We persuaded David Lee at Pulford Forestry to let us put Ginger back in the field at Whinfell temporarily, but we could not afford to pay him rent permanently. We had made a good start, but our position was tenuous and precarious. By following our noses, and helped by generous friends and more than our share of synchronicity, we had landed on our feet and were now tuned in to our new world, but we needed money and we needed grass. To get it we had to find more horse work, and soon.

The exhilaration of a fast learning curve had become a roller-coaster, in which each breakthrough was followed by another obstacle, but now we had momentum. St. Croix Wood was the turning point. After St. Croix we were committed. Going back was not an option.

8. Joss Rawlings - *apprenticeship*

St. Croix made me realise that I needed an experienced teacher. I knew almost nothing about timber prices, rates of pay, potential employers, woodland weights and measures, specifications and products, estimates and job costing, tree species, and thinning regimes. My time at the Arvon Foundation had showed me the value of apprenticeship and of the years of experience that distinguish the master from the novice. Once again I was uncannily lucky – I found one of the very few remaining horsemen in England who could teach me what I needed to know. I found him less than 20 miles away, near Greystoke, working for the Forestry Commission. His name was Joss Rawlings.

Joss was related to our immediate neighbours at Littlebeck, the Aireys. We had called on Mrs. Airey to buy eggs, and when we told her that were working a timber horse she told me about her cousin Joss. Shortly afterwards I visited the Forestry department library at Newton Rigg, the agricultural college near Penrith, in search of information about horse logging. Not only did they have half a dozen books on different aspects of the craft, but the helpful librarian found me a set of *Forestry Commission Standard Times for Horse Extraction*. This was a vital missing part of the jigsaw. The *Standard Times* converted the variable and almost meaningless unit "One pole at the roadside" into the precise and useful unit "One hour," by a series of calculations based on time and motion studies in different sizes of timber in different terrains. Those books were essential food for a hungry man - at last I had a firm basis for costing, pricing, budgeting, and estimating. After digesting the statistical information they contained, I could at last (in theory at least) negotiate on fair terms with timber merchants and timber owners. Although the *Standard Times* were never, ever, exactly accurate, that was because no two jobs were ever the same, and because in 1978 the Forestry Commission was still using a pay rate which had been fixed in 1963. In reality there was no such thing as a 'standard' time, but now at least I had a common language and a rule-of-thumb which allowed me to do proper job costing and make some plans with confidence.

The Newton Rigg librarian knew all about Joss Rawlings. The forestry students would spend a few days learning about felling and extraction at Greystoke Forest, where Joss worked. The librarian gave me the name of the Greystoke Head Forester, so we borrowed the Morris 1000 van and went to talk to the men at the sharp end of the business.

Juss Rawlings with his timber horses, about 1965

We had been warned that Joss started work early, and even though we were there for 8.00 am, he had already left his base and was somewhere out in the woods. We had been given rough directions for finding him in the 4,000 acre forest, and so for the first time we entered another strange new world – the deep, dark maze of a large scale commercial forestry plantation. The hardened forest roads were well engineered in a grid pattern, made to carry 30-ton lorries in all weathers, and every few hundred yards along the roads there would be a long straight grassy gap between the trees. These long gaps are known as 'rides' and are used as firebreaks, or as tractor roads, and they mark the edge of each 'compartment', as the different areas of forest are called. Each compartment would be planted with a suitable tree species, depending on the soil type, the slope of the ground, and the drainage. The common species were Sitka spruce, Norway spruce, European and Japanese larch, and Scots pine, with occasionally some Douglas fir, and only very rarely a small stand of native hardwood species such as oak, beech or ash.

As we drove past each ride we paused to look along it to see if we could see horses and soon spotted a fine great coloured cob, head down and grazing loose, with no head-collar or tether. Round the next corner was a heap of fresh timber and a newly made rough track out of the wood. We heard chains jangling and dry branches crackling, then another great coloured Clydesdale-cross cob appeared, pulling a mighty pole. Without giving us a look, and with no command and nobody else in sight, the great horse turned the load onto the road and stopped exactly at the end of the stack, again with no command. Joss appeared out of the wood a few minutes later, walking slowly and breathing heavily.

He was half expecting us, and was polite if not exactly friendly. We hardly got a word out of him on the first visit, and certainly not anything approaching a smile, while we explained our mission. I told him that I had been a farm horseman, and we were just starting out in forestry. We had a horse and a chainsaw and we wanted to learn the trade. Would he be willing to teach me? He was non-committal. He would think about it. Come back in a week.

So we left, a little downhearted that Joss had not been more friendly, but excited to find what we had been looking for since leaving Geoff Morton – the sights, sounds, horse smells and unmistakable expertise of an experienced horseman working at the trade we wanted. We could not understand why he did not jump at the chance of free labour, but guessed that he was anxious we might take up too much of his time, or even become competition to his business. Most forestry contracts are awarded by competitive tender, and there was only room for a limited number of contractors, so maybe he was reluctant to encourage us too much.

On the way out of the wood we called at the Forest Office and talked to the head man. Once again we explained what we wanted, and asked if he had any openings for another horseman. He was immediately interested, and we suspected that he wanted to hire us in competition with Joss, so as to keep the extraction price down. I told him that we would not want to be in competition until we had learned the trade, but if Joss agreed to teach us, would he be willing to let us move a caravan into the wood so that we could live on site? He agreed, and we arranged to see him again in a week and take it further.

A week later we went back to see Joss. This time we got there earlier – 7.30 am, and there he was, sitting in his car listening to the radio, eating the first of his many sandwiches, while his two great Clydesdales munched their hay. His main workhorse, Charlie, was 17 hands high, and steady as a rock. One of Charlie's knees was swollen and puffy from years of heavy straining and pulling, but had no heat in it, and it did not seem to bother him, although he moved carefully as if his joints were sore and every ounce of energy so precious it should not be wasted. When he was working, Charlie would lean into his collar to take up his load in slow-motion, and once he got the load moving he kept a slow and steady pace, like a plough horse. He would stop pulling whenever he wanted to take a breath, and wait for Joss to catch up.

Joss was short – about 5' 3", and almost as wide. He had a rolling gait like a seaman, due to bad arthritis and years of tramping over tree brash and rough ground. He controlled his horse with voice commands alone, following Charlie slowly down the snigging track, then coming up alongside the load to check that all was OK, giving a little grunt, and in a quiet breathy voice, almost a murmur, he would send him on. "Chullie, go on." Charlie would lean slowly into the collar and set off for another 50-yards pull, before he stopped for another breather. They had worked together like this day in and day out for 16 years, they knew each other well, and they were both growing weary.

Joss depended on Charlie for every crust he earned, and because he was worried that the old horse was failing, he had gone to Wigton horse sale and bought a new young horse. Sam was a young and green Skewbald gelding, a few inches shorter than Charlie, but much broader. Probably part Clydesdale (and the other part probably rhinoceros) he had a chest like a barn door, and a back so broad that I could barely get my knees apart far enough to ride him. He had an appetite to match, and although his temperament was not vicious, he was unbroken, unschooled, and already set in his ways, which were uncooperative. Within a day we called him Sam Pig, and within four months he had put me in hospital.

On our second visit, Joss was more friendly. Whether he had made enquiries about us with the Aireys, whether he had a change of heart about the threat we represented, or whether he had simply decided that working together might be to his advantage, we did not know, but we had a proper discussion. Yes, he would teach us, and in return he wanted me to break in Sam Pig to the timber job. Yes, I could use my own horse Ginger as well, because he could only afford to pay me what I earned, and Sam would not earn much for at least a month. I would provide my own hard feed, my own hay, and my own tools. I would work for Joss and not for the Forestry Commission. All fair enough. I agreed. We arranged that I would start as soon as possible. Game on.

First we had to find wheels of our own, and somewhere to stay. Next stop, Hebden Bridge, and Michael Phillips, our mechanics guru, who knew where there was an Austin van for sale. We hitched down to Hebden and went to find its owner, Jerry Whitehouse. We knew Jerry as Herr Idelbouger, which was the name he used when signing himself into commercial hotels in Holland. He took van loads of furniture, bought for a few pounds in the auction rooms of Northern England, and sold them on the street markets in Holland and Germany at a good mark-up.

In his spare time he went hang-gliding and had a rock'n'roll band. His Austin van had been in a collision and was ready for the scrapper – the back door, the back light and the back wheel were all damaged. With the help of Michael Phillips, the walking mechanical encyclopaedia, we hammered the dent out of the door, replaced the damaged wheel, blew up the tyres, and with gaffer tape we fixed a rear light cluster taken from a scrap lorry into the hole in the panelling where the van rear light should have been. Since it had no MOT or road tax we had to avoid the Police patrol cars, but we managed to get it all the way home to Cumbria by keeping off the A roads and sticking to the tiny back lanes between Bacup and Kings Meaburn, most of which we already knew from our horse-drawn trips to Appleby.

Once the van was legal, we set off to Whitworth again to see Stuart Nutt. Stuart was a travelling man, a general dealer and scrap man, and we thought he might know where we could buy a cheap trailer caravan. We found out where he was stopping, pulled into the traveller site, and told him what we wanted. 'That one is for sale' he said, pointing across the yard at a 22-foot Bluebird living van, made in the late 1950s. It had a fold-down bed, a wood-burner, pink mirrors, and a separate kitchen. Gas lights inside, no running water, no spare wheel, no jockey wheel, no brakes, and no road lights. "How much?" "£150." I knew it was too much, and it turned out it was not even his to sell, but he made sure he asked us a price that allowed him buy and sell it on at a profit. Anyway, we had few options. "Find me a spare wheel, and pull it up to Penrith, and it's a deal." "OK, but you have to pay the diesel." "Done!"

And so, on 14th January 1978, we had a draught horse, an Austin van, a Bluebird caravan and a debt of £600.00. We set off back to Cumbria in high spirits. The dodgy Austin van had picked up a puncture on its barely legal tyre, and it had no spare wheel, so we loaded it on the back of Stuart's flat-bed truck and rode in the cab with him, while he towed the new caravan. That was an expensive mistake. We arrived at Littlebeck, unhitched the caravan on the road outside the house, paid Stuart his diesel money, and looked around for ramps to unload the van. While we were looking, Stuart was in a hurry to get back so decided he would use his crane to unload. He fixed his chains, lifted the van off the flat-bed, and as we came round the corner we saw the van suspended precariously. As we watched, the van tipped, the chain slipped off the front, and it dropped square on its nose from a height of about eight feet. Stuart laughed – 'Oh, I dropped it' was all he said. It would never run again. Stuart was back in his cab and out of there within about a minute. We are still friends, and he has more than made it up to me since then, but it was not a great start.

The mishap forced our hand, and within a week we had a better vehicle. Our friends Pete and Mel Landells changed their old Morris 1000 Traveller for a not-quite-so-old BMW, and we bought the Morris for £30.00. It lasted well, and by the time I terminally unzipped the main chassis member on a protruding tree root two years later, I had done almost every mechanical job in the manual. But that all came later; for now, we were all set.

Greystoke Forest 1978
Morris 1000 Traveller, caravan, bender tent (right) and Ginger.

Within a week, we had our 22-foot bedraggled Bluebird towed from Littlebeck into Greystoke Forest, and we parked it down a hard ride about 300 yards from the main gate. We went back for Ginger, boxed him to Greystoke, and turned him into the enclosure with Joss's horses. We cut two dozen hazel rods, attached the ends to a ridge board, stuck the other ends in the ground, and slung a canvas wagon sheet over the lot to make a traditional Gypsy bender tent. There we kept the harness, tools, hay and horse nuts, chainsaws and fuel, spare gas, wet clothes, and an Elsan chemical toilet. We spent a day collecting firewood, sorting out and settling in, and then we started work.

Joss was a good teacher. He watched me working, put me right where necessary, and suggested different ways to do things – 'the wrinkles' as he called them. In particular he taught me how to shorten the main load chain with the ring so that it could be undone quickly without jamming and he showed me how to use the horse to build a stack of poles at the roadside, 10 feet high or more, with no crane, no winch, no skyline, and no cables. He taught me about the politics of working for the Forestry Commission, and how to try and stay on top when they were trying to squeeze you down on price using Standard Times that were close to fiction. He taught me about how very quickly timber loses weight as it dries out, and how the merchant would delay sending the wagon if he was paying the horseman by weight, when he himself was being paid by volume. He taught me about hoppus measure (also known as quarter-girth, the 'old way' of measuring round timber) and when we eventually parted company, he gave me his spare hoppus tape as a parting gift.

Over the next few months, while sitting in his car waiting for the rain to stop, Joss told me the story of his life, which was uncomplicated. He had started as a forestry horseman on the Dodd, near Keswick, in 1932, and he had never done anything else for over 45 years. The Dodd was notoriously dangerous – a steep and difficult slope overlooking Bass Lake, and Joss had known of three horses and two men killed while taking timber off that slope. Joss had married, and had lived with his wife in an old barn for the last 20 years. He never mentioned any children, so I never knew about his family. Joss was close to retiring age when I met him, and struggling with diabetes and arthritis. I would arrive at his base camp at 7.45 am, to find him standing next to the electric fence which he used to corral the horses overnight. He would be gripping the electric fence wire with a huge hand, and the whole of his arm jerked with each pulse of electricity. "Can't you feel that, Joss?" I asked, amazed. "Oooh aye, I can," he said "… reet down to me ankles." He reckoned it was good for his arthritis. I didn't argue, but I didn't try it.

Once the horses had cleaned up their feed, Joss would open the back of his Austin Maxi where he kept his harness, and yoke up Charlie for work. Using a long rein made of sash cord, he would drive the Maxi slowly down the track into the deep dark wood, with his long lead rein passing through the open window and Charlie plodding on behind, while I rode in the passenger seat doing the same with Ginger. Joss ignored Sam Pig, who would hang about and refuse to follow until we were almost out of sight, and then come galloping after. Sam would wander about looking for fresh grass while Charlie was working, or else stand rubbing his fat backside on a tree, but he was never far away at bait time.

This strange slow procession into the forest became a memorable part of our routine, a calm prelude to the adrenalin and excitement of days full of chainsaws and hauling timber. So began my apprenticeship.

Ali, long-reining a sawlog with Ginger, Greystoke Forest 1978.

9. Greystoke Forest - commercial silviculture

Greystoke Forest covered 4,000 coniferous acres west of Penrith, and it swallowed us up for the winter of 1978, so that Ali and I began to feel that we belonged in the wood. We detected a certain reserve among the woodmen, a tightly-knit bunch who had worked together for longer than we had been alive. Not only were we young, but we were not local, we did not speak Westmorland dialect, and we were novices, but we were treated with courtesy and respect. Some men would tell us nothing, because then we would know as much as they did, and some men would tell us everything they knew, because then we would know as much as they did.

We settled in to a routine, and there is nothing like a routine to absorb the elements of a strange and different culture. We would work in light rain, but not in heavy rain. After 40 years of working in the woods, Joss had a refined sense of what the weather would do. At any time of day, if there was a heavy shower he would know whether it would pass and we should sit it out, or if the day would be ruined and so we should go home. I worked with him for about six months, and he seldom got it wrong. Sure he listened to the forecasts, but they were wrong more often than he was. When the rain came and he thought it would pass, we would sit in his car and listen to Radio Cumbria. The 'Swapshop' provided our regular entertainment at bait time (*'Lady in Workington with parrot cage would like to swap it for an electric toaster'*...etc.). Sometimes in winter he would put a canvas sheet over Charlie if he was sweating in the cold, but mostly the horses just stood in the rain, chomping, while Joss answered every question I asked him, or he told stories of his days as a farm horseman in the wood. He told how he had ploughed, harrowed and rolled 'till his arse fell off', using a wooden plough on the stubble. His favourite job was driving geese to King's Meaburn at Christmas, and taking a dram of whiskey at every farmhouse on the way. He once lost a pocket watch after hanging it on a branch while he was working and then forgetting which tree he left it on. Twenty five years later the watch appeared, buried inside the trunk when the tree was cut up in the saw mill.

After a few weeks we were getting the hang of the job and producing almost as much as Joss and Charlie. My first wage from Joss was £43.00, and to celebrate I declared my existence to the Inland Revenue and registered as self-employed, even though my wage was too low to sustain us for very long. Although most of the production work in Greystoke Forest was carried out by self-employed contractors, the Commission maintained a gang of half a dozen men who lived in the forestry houses at the edge of the wood. The foreman was Cuthbert, a small wiry man with a big smile and a floppy woolly hat, but short of one eye and part of an arm, including several fingers, due to his chosen occupation. In my seven years in the wood I would regularly meet woodmen with body parts missing – usually the fingers. Some were short of toes, and many had bad backs, but all made light of their injury, and carried on. "Well, you have to do, don't you?" was as much as they would say about it.

Neither Joss nor Cuthbert's gang were keen on 'converting' the timber we extracted. The job was to cross-cut the roadside heap of poles into specified lengths, and stack them ready for the timber lorry, so it was simple enough but it was heavy piecework and poorly paid. I was young and keen and strong, I needed to earn more, and work was work, so I arranged with the boss that I would do the conversion.

The usual specification was a combination of 1 metre pit props for the coal mines, 2- or 3-metre pulp for Bowater's paper mill, 7 foot TBM, (Thames Board Mill, a precursor of MDF) and sawlogs for the sawmill. Anything bigger than a nine-inch top diameter and four metres long was a sawlog. Anything less than a four-inch top was a fence post, up to 5' 6" long. Anything less than a two-inch top was scrap and would end up as firewood. I was paid piecework at a rate of £2.00 per ton. It was proper graft, and too much of that £2.00 per ton went on keeping the chainsaw fuelled and sharpened, but I could work an extra hour or two after Joss finished at 4.00 p.m. and make another £6 per day, although I had to move three tons of wood to do it.

Ali did not work much with Ginger, but she was never idle. She helped a neighbour in her garden, after the lady asked "Can you stitch?" meaning "can you grow potatoes". She made all the meals, wrote letters, and kept the caravan as cosy as we could wish, but that did not occupy all her days, so she asked the head forester if there was any paid work she might do. He suggested brashing, which turned out to be a hard and dirty job, but she agreed to give it a try.

Brashing spruce trees is an exhausting and tedious job. Commercial plantings of spruce are almost impenetrable after about six years growth – they are planted at a density of up to 1,500 trees to the acre, and after six years the side branches have grown out to form a dense, prickly and unyielding wall. It is impossible to move among the standing crop, which soon becomes a dark, damp and perfectly unattractive world of its own. In order to move among the trees to inspect them, and to see where they have failed to grow (known as 'checked areas' – usually bogs) or to mark them for selective thinning, it is necessary to 'brash' the trees by cutting all the side branches off each stem, up to a height of about six feet, using a curved saw on a long handle. A series of sharp downward cuts is made, close up against the trunk, removing a dozen or more side branches. Some would be wispy twigs, but others were twice the thickness of a thumb, taking some serious effort to cut, and some of those were a foot above Ali's head.

She was provided with a brashing saw and a set of 'standard times' which set the rate of pay - about 2p per tree. After about two hours of exhausting work, she had earned, according to the book, about £1.50. No good at all. She marched up to the office to re-negotiate. "Can't you make it?" said the Forester "No? I didn't think you would" he said - "you need to work quicker." "Not possible." she replied. "Come on, I'll show you how" he said. He dived into the wood to give a demonstration and went at it like fury. The snow fell off the branches and went down his neck, his jacket snagged on the spikes, his wellies filled with sawdust, his face grew red and his breath got short. After an hour he retreated to his cosy office, having proved it possible to earn about £2.00 an hour, at 3p a tree, but he was exhausted and there was not a chance he could keep up that pace for an eight-hour day. He offered a 50% rate increase, but even that was inadequate, and it was clear there was no future in it for Ali. After a month she had brashed 1,200 trees at 3p a tree, and earned all of £34.20 after deductions, so then she reckoned she had better things to do.

Luckily we had few chances to spend what little money we had, so even on our low wage, at Greystoke with Joss we were able to start to pay off our debt, slowly, and still maintain a basic social life. The local pub was a walk away, and a short drive away was one of the last pubs of its kind, where there was no bar - just a room full of chairs around the wall, occupied by old codgers playing dominoes for hour after hour, holding them close, and scrutinising them intently while chatting about nothing in particular. The landlady would bring the beer in a jug from a back room. We loved the old world atmosphere, and the constant banter, but could not afford to be regular drinkers. We did like to dance though, and once a month we drove 15 miles to Ireby for the Ceilidh dance, where we whirled and twirled into the night to the music of the exhilarating Ellen Valley Band.

One weekend we climbed Carrock Fell, where the Romans mined lead, and where the wild juniper bushes clothe the steep slopes. Once a week, on market day, Ali would drive to Penrith to do the laundry and buy our supplies. The standing order was a bag of flour, a big bag of vegetables, half a pound of mince, a pound of bacon, half a pound of

Cumberland sausage, and half an ounce of Virginia tobacco, plus a tank of petrol. For all that we had change out of a £10 note. Otherwise, we ate mostly wholefoods bought in bulk - brown rice, lentils, bulgur wheat, miso sauce, wholemeal pasta, and porridge oats. Ali prepared delights taken from *The Pauper's Cookbook* - baked red-cabbage & onion, and belly-pork & beans - cooked on the ancient gas burner. The wood stove was fired up with fence post pointings to achieve a sauna-like atmosphere while we sat in our T shirts and dried out the wet work clothes.

One bonus came our way in the form of a job that only a horse could do - taking down hung-up trees in selective thinning. The Commission woodcutters were working in a stand of overgrown Sitka spruce, trying to do a first thinning of trees 35 feet high and six feet apart. They would cut out the 'gob' at the base as normal, but when the final felling cut was made, the tree would go nowhere, but just stand there, supported by the dense canopy on either side. These were known as 'hung-up' trees, and the usual way to get them down was to use a cant hook, or what was is now known as a turning bar. The cant hook is a long wooden shaft, up to 6 feet long for a big one, with a hinged sharp metal hook on an arm at one end. The technique is to place the bar parallel with the ground, and dig the sharp hook into the butt of the tree. By pulling or pushing the end of the bar, the tree would be twisted on its vertical axis, and would usually roll off the stump. As it rolled it would lean a little and twist its way through the canopy and onto the ground.

The problem for our woodcutters was that these hung-up trees were so big that, as they twisted off the stump, they would stay vertical and dig in to the wet ground, stuck fast. No amount of cant hook work or brute force heaving could move them. A few trees hung-up like this could be managed by felling the neighbouring trees which were causing the hang up, but in an overgrown plantation almost every tree had the same problem, so taking out the extra trees to make the felling easier would sacrifice better trees that should be left to grow on, and more importantly it would open up wide holes in the canopy and let the wind in. Before long the whole plantation would be windblown.

Windblow not only causes significant timber losses, but is the most dangerous of all working forestry environments. When windblown trees are uprooted, they lie under severe tension or severe compression, often in a gigantic tangle, like the old game of Spillikins. To harvest them, each cut with the saw must be made with extreme care, after weighing up the stresses, and predicting the effect of the cut on the whole tangled mass. If you get it wrong, a four ton tree can move along the ground at a speed faster than the eye can follow, and serious or fatal injury is a constant hazard.

Even a windblown tree lying flat on the ground can be dangerous. When a windblown tree lies more or less horizontally, the roots rise up out of the ground to form a vertical plate, 8 to 10 feet high, held upright by the weight of its own fallen trunk. When the tree is cut off at the base, the root plate is released and falls back to where it came from. Sometimes it falls back immediately, sometimes it never falls back if the tree has been windblown downhill, and sometimes it stands upright for months and falls back unexpectedly. Whenever it falls, if there is someone underneath the plate at the time, they die. While we were working in Greystoke a man was killed this way in a Forestry Commission wood in the south country, and we heard about it on the grapevine the same day. The man had been sheltering from the rain behind a cut-off root plate. It had been standing for a few days, but it had been inching back slowly over that time, until it over-balanced and fell.

Windblow represents a major loss to the forest enterprise, and opening the canopy always increases the risk of windblow, so if possible any hung-up trees need to be taken down without felling the trees all around. The hang-ups we had to deal with were third thinnings, too heavy for a man to lift, and because it was selective thinning, the tractor could not get into the wood among the standing trees, so the horses were sent for.

We had to agree a price, and Joss asked for 60p a tree. This was not agreed, so he walked away. Next day the boss came to see us to negotiate. Joss was immovable, and 60p a tree was agreed. We walked the horses about a mile to the 'Berrier end' of the forest, and followed a four-man cutting gang into the wood. The technique was to wrap a long snigging chain a few times round the butt, just above the cut, while it was still resting on the stock. As the horse pulled the butt away and off the stock, the whole tree would twist and spin itself out of the canopy as it fell. If it dug into the ground the horse would just keep pulling, and the wrapped around chain would spin it out.

We each pulled 100 hung-up trees that day, and it counted as fairly light work for us and the horses. We collected £60.00 each; almost double our theoretical day rate. I asked Joss how he managed to get such a good rate "No problem - he came like a lamb. He had no choice." Joss knew the price of the cheapest alternative, which was to hire an elaborate portable winch, with operators, and carry it in to the wood. Joss priced the job just below the cost of the winch. I was beginning to learn 'the wrinkles.'

Greystoke Forest (Photo: Peter McDermott)

Apart from the occasional bonus, we were living at subsistence level, but life was good enough. Our pleasures were simple - we watched the goldcrests and red squirrels high in the trees, we watched a family of foxes playing on the green ride, 20 yards from our caravan, and a family of roe deer that bounced through the clearings, their white tails bobbing like ping-pong balls. I had an ancient Uher reel-to-reel tape recorder and an even more ancient Roberts valve radio, both powered by a car battery, and we listened to tapes of *The Folk-Music Virtuoso* and to *The Hitch-hiker's*

Guide to the Galaxy on the radio. In the evenings we played Irish tunes for hours on the whistle and fiddle, and now and again we would wander home mildly drunk from the pub in Hutton Roof, or take a day off to go to a farm sale or a ploughing match.

At the weekend we might find an Irish music session, or drive 50 miles to see our friends in Wensleydale, where we would drink home brew, roast a leg of lamb, sing songs and play poker till the small hours. On the way home we would call in at the rope maker in Hawes to buy halters, long reins or galvanised shackles, and then to the Hawes grocers where we would order 2lbs of Wensleydale cheese, just for the pleasure of seeing them unwrap a whole cheese the size of a motor tyre, and cut our piece out of it. Ginger was working well, we were fit and healthy, and living our dream. Spring was coming. What could go possibly wrong?

After about a month with Joss, he began to drop hints about the young horse, Sam Pig. He wanted him broken in to snigging, partly because it was well overdue, (Sam was rising seven) so that he had a reserve horse if Charlie got too old, and partly so that he could work two horses at once. Much of the time taken by horse extraction is the travelling to and fro in and out of the wood. If you go in with two horses and come out with two horses, you don't quite double the output, but you can produce about half as much again as with a single horse. At that time a standard rate for a man and a horse was £30.00 per day, and with an extra horse you could make it £45.00, so it was a sensible move. It was also part of the deal I had with Joss, so it had to happen.

We arranged for the farrier to come and get Sam shod, and next day I yoked him to a sawlog and tried him out on the forest roads at the end of the work day. We started with a small load, but he did not like it much, and because he had never been properly 'mouthed' I had to use a lot of strength to hold him back on the bit. Not a good way to start, but we persevered and day by day we increased the weight and the length of the pull, and he steadied down a little. After a week he could pull a decent load for a few hundred yards without bolting and after three weeks I decided to try him in the wood.

At that time we were working on first thinnings, so a full load was not single pole but could be up to twenty small poles, usually about 15 feet long, bound together with the snigging chain and pulled out as a bunch. There was a high proportion of brash and a bit of an uphill jag (a Joss word for a short, steep slope) to get from the edge of the wood onto the forest road, which was built up on a causeway through the bog. Uphill dragging through brash is seriously hard work for a horse so I could not load Sam too heavily but we managed a few loads, and took a break.

Sam was a complete novice, so could not work to voice commands, and there was not room among the first thinnings to work him on long reins (known as 'cords') so I would bring him out of the wood on a lead-rein – a short rope fastened to a ring on the head-collar. With a lead rein, the standard way to go forward with a horse on a road is to hold the rein in the right hand, with the horse on the right side. Because it was vital to watch the load at the same time, so as to avoid any of the poles from jamming up, and because I could not be looking over my shoulder all of the time, I would walk backwards, with the lead rein in my left hand, and with the horse on my left, glancing over my shoulder to watch where we were going. As we came up the jag onto the road, we had to turn hard left and drag the load up onto the stack, tip first, so that the stack lay parallel with the road. The first few loads worked well, and we had a decent looking stack, but on the first load after our break, Sam was jumpy. As I pulled his head round to the left to make our turn at the top of the jag, he was moving at speed, and I was watching the load come up onto the road. As he turned, he trod on my left foot.

He weighed a little under a ton, and he was pulling about six cwt uphill at about three miles an hour. I never bothered calculating the inertia or the downward pressure on my foot, and luckily I was wearing steel toecaps, so nothing was broken, but I could not move my foot while he was treading on it, so I could not get out of his way, and he knocked me over. I tried to hold on to the lead rein (another mistake) but by the time he had lifted his front foot off me, I had already fallen backwards to the ground, and he trod on my leg with his back foot, and then dragged part of the load over me, before setting off down the road to his feed station, where, luckily, he stopped.

I was a bloody mess, and so badly bruised and battered that I could barely stand, and could not walk. I crawled into the grass and lay there for a while, wondering what to do. Joss and Charlie were working in another compartment so as not to distract the young horse, but even though I needed help I was glad he was not there to see me. I stood up, and with a combination of a limp and hop for half a mile, was able to get home, bleeding and lame, and wondering how much damage was done. Ali cut off my jeans, cleaned up the wounds with Dettol, and drove me straight to Carlisle Hospital. I was x-rayed, checked for reflexes and concussion, bound up, jabbed for tetanus, and sent home to lick my wounds.

When going over in my mind the sequence of events to work out what exactly had gone wrong, I remembered that Joss had showed me a particular technique which I had forgotten to use. I never forgot it again. When walking backwards, watching the load, with the horse on your left hand side walking forwards and holding the lead rein in your left hand, you use your spare right hand to maintain a safe distance away from the horse's chest. You do this by stretching out your right arm, locking the elbow, and placing the flat of your right hand against the horse's shoulder, in effect walking sideways, not backwards. As the horse moves forward against your locked right arm, he pushes you back, and away from his great big numb iron-shod clomping feet, so he cannot tread on your toes and knock you over. I never forgot that one ever again, and even now if I have to lead a horse while watching the load behind, I automatically stiffen my arm and keep my distance.

It took two weeks for me to recover, but after a couple of days I was up and about and began to take stock of our position. My cuts and bruises healed, but my confidence in the grand plan had taken a hard knock. Clearly horsepower was not plain sailing and Mother Nature had a nasty bite, but although I was defeated and damaged, we were by then so far committed that we did not seriously consider giving up. I blamed myself and my bad luck and inexperience, and I certainly did not fancy going back to handling Sam Pig. I had so little time in the day after trying to earn a living, that it was a hopeless task anyway. I knew what Sam needed, and I told Joss. He needed to be boxed down to Morton's farm in East

Yorkshire, and yoked up as the fifth horse in a five-horse, three-furrow ploughing team made up of seasoned Shires and Clydesdales. He needed to be a little fish in a big pond for a change, and not the other way round. Sam was big and powerful, but compared to a seasoned Shire plough horse he was a lightweight. Once a big ploughing team got moving he would have no choice but to steady down and do some serious work – if he did not find their rhythm and work alongside them straight away, he would simply be dragged along by the power of the team, which would hardly even notice one pig-headed young wood-horse. If he pulled too hard he would tire himself out in five minutes and have to slow down. Once he had worked a full season before the plough he would learn the job of a draught horse, and might then make a good timber horse. There was no way that I could achieve that with a few hours a week pulling logs around. There was barely a log in the forest that he would not pull for fun, and anyway I had to get back to earning a living.

I broke the news to Joss that there was nothing more I could do with Sam, and he grunted in agreement, which I took to mean that he knew that all along.

10. National Trust - *amenity silviculture*

During one of our sessions sitting in the car in the rain, I had asked Joss if he knew anyone else working horses in the Lake District. He knew of one only, Harvey Bowe of Keswick, who worked for the National Trust, but was about to retire. I stored this information away, and while I was recovering after my accident with Sam, Ali and I decided that it was time to strike out on our own, so we made a trip to see some possible employers.

First we drove to see the Forestry Commission at Grizedale Forest, near Hawkshead, to check out the potential for work, and ask about the mysterious Norwegian skidding arches, which I knew about but had never seen. The skidding arches and harness were in immaculate unused condition, some still wrapped in corrugated paper from the factory, and the Forester was willing to sell them, but at a price which included a number greater than one, followed by three zeroes. They had lain there unused for years, they would lie there for years to come, and he knew very well that we could only earn our wages according to the Forestry Commission Standard Times so were living hand to mouth, but he would not budge. We said goodbye politely, cursed him bitterly when out of earshot, then made an appointment to visit the Woodland Department of the North West Region of the National Trust, hoping for a better reception.

Ken Parker, their Head Forester, was a diamond. He worked out of the National Trust Regional Office, at Borrans Road, in Ambleside, received us courteously, and listened to our story. I did not want to tread on Joss Rawling's toes by competing with him, as there was not enough work for both of us in Greystoke, so I needed my own position. I mentioned that I had heard that Harvey Bowe was about to retire, and wondered if there might be an opening for us. Ken Parker was immediately keen. He did not want us to work in Keswick where Harvey Bowe had been, but at Parkamoor Wood on the east side of Coniston Water. He favoured horse extraction because not only did he manage hundreds of acres of woodland on steep and rocky terrain, which was only workable by horses, but he was keen on selective thinning, where horses were more efficient and did significantly less extraction damage than tractors. The fact that we were mobile, self-contained and ready to go was a bonus for him.

Most of the woodland owned by the National Trust was traditional broadleaf, now amenity woodland, planted for horse extraction long before tractors existed and difficult to work with modern machinery. The Trust had a gleaming new Norse winch stored in their yard at Boon Crag, near Coniston, but they had not yet trained their woodmen to use it. The Norse winch was a tractor-mounted three-point-linkage drum winch, with multiple 'choker' cables allowing half a dozen poles to be attached to the line and drawn up to the drum before being lifted off the ground on the hydraulics and dragged out of the wood. It could be fitted to a standard tractor, although forestry tractors were usually modified with guards to protect the driver, the radiator and the tyres. The Norse winch had become the small woodland owners' preferred extraction method for first thinnings, but it was only really effective on fairly level ground in line thinning - even the Norse winch could not compete with the horse for selective first thinning on uneven or steep terrain. Unfortunately, by the time we arrived in the wood, most commercial woodland owners (as opposed to amenity woodland owners such as the National Trust) had recently abandoned selective thinning in favour of line thinning, because it was cheaper, due in part the technological advance of the Norse winch, which had become standard equipment.

By the time we talked our way into working for the National Trust, my crash course in 'Teach Yourself Silviculture' had provided me with some basics of forestry, and at that time the theory and practice of thinning was fundamental. In order to produce high quality sawlogs. 1,500 trees per acre would be planted close together and the plantation would have one, two or three thinnings before it was clear felled. The thinning concentrated the volume of timber into a smaller number of the best trees. Alternatively, for maximum-volume, low-cost, low-quality production intended for making chipboard or fibre board, the trees would not be thinned out at all before the wood was clear felled and replanted.

In selective first thinning, the low-quality suppressed trees (short, spindly and/or dead), together with the wolf trees (distorted, bent, forked or twisted) and any obviously overcrowded stems would be selected and marked for felling, leaving the healthy and dominant intermediate trees to grow on. In the second and then third selective thinning, the weaker or smaller trees would again be progressively removed, leaving only the dominant trees, which would grow into the spaces in the canopy created by the thinning. This process of selective thinning produced a high proportion of high quality sawlogs - straight, clean and free from side-branches and knots. For the best quality timber, a further process of high pruning was carried out, using a heavy brashing saw on a long pole, so that the resulting sawlogs would have 12-15 feet of knot-free wood.

First and second selective thinning does not usually make any money for the forester. In fact it costs money, because the value of the thinnings produced is less than the cost of harvesting them. If the objective of the plantation is to produce the maximum volume of high quality timber then first and second thinnings are usually considered part of the establishment cost, although third thinnings in a high yielding wood might break even. If the quality of the timber is important, then selective thinning is essential.. The two great expansions of Forestry Commission planting occurred after the major world wars, when the strategic objective was to produce a national reserve of quality timber for a future national emergency such as a war effort. (The U-Boats exposed the vulnerability of imported goods in wartime, but one of our most successful military combat aircraft, the Mosquito, known as 'The Wooden Wonder' could be constructed of plywood made from clean, straight-grained Spruce trees.) Quality sawlogs were required as a national strategic reserve, and the Forestry Commission was given the job of planting and maintaining that reserve to be available if needed.

Horse, horseman and chainsaw

Twenty years later, when the plantations were ready for thinning, they went to feed the pulp mills which produced flat, even sheets of processed wood fibres, made up of lignin and cellulose macro-molecules. When high volumes of quality sawlogs for milling into planks were no longer needed there was no longer any reason to do any thinning at all. Silvicultural research had shown that in a 'closed canopy' (i.e. where the tops of the trees had all grown close together, blocking the light that reaches the forest floor) the total volume of lignin and cellulose molecules produced by an acre of land of any given yield class would be almost the same, whether it was produced by 100 trees or 1000 trees. For wood that was to be processed into paper pulp, or compound board such as TBM, or later MDF, it was more profitable for the forester not to do any costly thinning, but simply to clear fell each plantation as soon as the majority of the trees were mature and before they were overgrown or windblown. The clear fell would take everything at once, regardless of quality, and the whole lot would be sent for paper pulp or to the board mill. No more thinning, no more horses. No more horsemen. Enter the big machines.

As selective thinning faded out, so did the demand for horse-logging. I came to rely on conservation organisations such as the National Trust and the Lake District National Park, which recognised the benefits of high quality timber, and particularly the amenity and conservation benefits of thinned woodland with an open canopy. Un-thinned Sitka spruce woodland is classic monoculture, one of the most depressing and undiversified living environments on the planet. (Life, but not as we know it.) Very little light can penetrate, so the flora and fauna are almost entirely absent, apart from a few invertebrates and the occasional rodent that feeds on them. A few birds find a habitat there, but their food sources are limited; a few foxes feed on the rodents, and the forest rides can make sheltered grazing, so are used by roe deer, rabbits and badgers, but that is about all that can be said for the wildlife value. There are no cycle ways, no forest walks or nature trails, no bird-watching hides, in fact very little wildlife or species variety, just acres and acres of low grade timber to be turned into pulp. The pulp in turn goes for newsprint or for wood products that are almost certainly destined for landfill within an offensively short time scale. Yesterday's newspapers are seldom treasured, and there are few craftsmen who make quality heirlooms out of MDF.

My objection to the 'no thinning' and mechanised harvesting policies was the centre piece of my attempt to convince woodland owners that horse extraction was good business, but with hindsight I was never going to change the world of commercial forestry. The financial and productivity gains of mechanisation still outweigh the social, philosophical and ecological costs, but trees and fossil fuels are at the centre of the scientific and ideological debate about climate change and global warming, so my arguments are worth repeating 40 years later.

The business end of a tree harvesting machine.

The Luddite objection is simple enough. The tree combine-harvesters, which in 1978 could cut 90 cubic metres (about 78 tons) per hour, all untouched by human hand, had put many thousands of skilled woodmen, including horsemen, out of work. The Forestry Commission's figures reveal that it would take 24 chainsaw operators to process as much as one tree harvester, which burns 90 litres of fossil fuel per hour. I didn't like any of that, but I could hardly object on Luddite grounds alone – unemployment is a universal effect of mechanisation. It is hard to deny that mechanisation relieves humanity of drudgery, and some employment is created to manufacture and maintain the machines, so I was not a Luddite, and I had to concede that point at the time.

My second objection to mechanisation was based on the distorted accounting procedures which underpin it, and which are concerned only with short term cash flows, while the hidden costs and negative effects are ignored. Horse power is truly sustainable and easily renewable, with low environmental impact, less damage to trees, low carbon footprint, low pollution, low capital, and low depreciation. These considerations were trivial to the management accountants who ran large scale forest enterprises. When preparing a conventional set of forestry accounts, the high capital cost of machinery could be seen as a positive benefit in the business plan. There appeared to be no shortage of available capital seeking a suitable investment vehicle, and the high capital cost and high depreciation cost of machines could be offset, quite reasonably, against the profit created by their high output, thereby reducing the tax liability on the profits from woodland, which was in any case already free of capital gains tax. There are standard accountants' calculations whereby the cost of the machine scores a double-whammy. The cost of the capital to finance the purchase of the machine, expressed either as interest on borrowing or as the 'opportunity cost' of the investment, *and* the depreciation cost (or technically the capital allowances) can *both* be used on the same machine to offset tax liability on trading profit in the same set of accounts. These capital and income tax incentives have been offered by Her Majesty's Governments (on both right and left of the political spectrum) to encourage the profitability of private forest enterprise and help maintain national timber stocks. As a result, high output harvesting machines are an efficient way of turning 'surplus' capital into 'tax free' income, without the costs and liabilities of personnel and pensions.

Horses appear to be a much less efficient way of doing this, because some of the real cost of providing the tax incentives for mechanisation is hidden. The real cost of carbon pollution from manufacturing and operating machines, from built-in obsolescence, degradation of the rural economy through unemployment and the depletion of our limited natural resources of minerals and hydro-carbons all have costs which are simply not included in the investment calculation. The immediate benefits go into the balance sheets of the landowners in the form of tax relief and increased profit. The gains are privatised and the hidden costs and losses are socialised, nominally for the sake of the 'strategic reserve'. This financial chicanery was not (and is not) confined to forestry, but was (and is) a feature of most mechanised, centralised production systems, so even the 'hidden cost' argument would hardly justify my objection to mechanised harvesting on those grounds alone.

Although I lamented the loss of craftsmanship and quality timber, my objection to mechanised harvesting was not merely snobbery about the production of wood for MDF (Medium Density Fibreboard) or chipboard instead of quality timber for craftsmen to make fine furniture, although it is a fact that most MDF goes into landfill more quickly than does traditionally made furniture. According to the Furniture Industry Research Association (FIRA) the amount of MDF waste produced at the production and manufacture stage in the UK is in the region of 180,000 tonnes per annum. (2013 figures). MDF cannot easily be recycled to produce more MDF, so most of this waste, (almost 99%), is sent to landfill. I have not found figures for the volume of finished MDF product that falls apart and follows the production waste into landfill, but common sense and observation suggests that it is a high proportion of what is produced. I know from observation that waste skips are full of it, and with good reason - have you ever tried re-assembling or repairing MDF furniture after moving house?

Although I did not care for the tax-payer subsidised black deserts of Sitka spruce grown to produce MDF instead of good timber, I could hardly blame the death of craftsmanship on MDF, which is a product of the free market. I was reading Solzhenitsyn's *The Gulag Archipelago* at the time, and the only serious alternative to the free market seemed to be socialist systems, which had either imploded and sent millions to their deaths in labour camps (Russia and Cambodia), or stagnated (Cuba), or 'reformed' and continued the destruction of the planet's resources with highly mechanised ferocity (The People's Republic of China.) Craftsmanship is a rare commodity, whatever the political system.

The social cost of the shift to mechanised harvesting was high. Not only did the accountants' 'bottom-line' and the relentless drive for profit through commercial forestry wipe out a tradition of craft skills and silvicultural excellence, but it undervalued the maintenance of amenity woodland for recreation and landscape, inhibited wildlife conservation, and limited species diversity, while increasing our reliance on imported timber of quality. It is notable that although UK softwood timber production is now four times greater than it was in 1978,[1] most of the softwood we use is imported.[2] That raises a question about amenity and conservation policy in the overseas woodlands that produce these imports, but Forestry Certification is intended to salve our consciences on that topic.

Mechanisation and intensive production for board mills was the main cause of the abandonment of the highly successful and effective Forestry Commission social policy, which once provided a network of affordable smallholdings for forest workers. The Acland Report of 1916, which paved the way for The Forestry Act, identified large areas of upland Britain as depopulated 'waste', and proposed that they should be planted up with trees. These plantations would not only increase the productivity of the land, but they would 'demand a higher rural population' than sheep rearing. The report proposed that the smallholdings be grouped together on the best land within or near the forests, so that forestry workers could work the smallholdings and the smallholders could work the forests. The proposals were immediately adopted in a clear and progressive policy. At the peak in 1958, the Forestry Commission had 1511 forest workers' smallholdings, but this had fallen to 906 by 1964, and in the 1970s there was a drive by all governments to sell off 'surplus' land. The Thatcher administration encouraged tenants to buy at a generous discount, and that was the end of that. (Source: Hansard.)

Taken one at a time, each negative effect of mechanisation might have been tolerable, but taken together they seemed to me an inexplicable collective insanity, all in the service of unreconstructed capitalism and its poor blind servants, materialism and consumerism. The case for horses seemed overwhelming, yet the drive for mechanisation was relentless.

---

1 *FC Wood production: England, 1976-2011*

2 *FC Timber Utilisation Report 2011*

In spite of my logical and rational objections, I could see why we lost the arguments. I had no wish to end up like the heroic steel-driving man John Henry, who tried to defeat the steam-drill and died with his hammer in his hand. Mechanical harvesting is safer and less arduous than traditional methods, although in exchange for the dangers of manual labour we now have obesity and risk aversion. (I know which I prefer.) MDF allows more people to have cheaper new houses, more affordable

Extracting windblow at Thornthwaite
Picture: Forestry Commission 1964

furniture, replacement kitchen units, and fancy staircases, which are all perfectly reasonable aspirations. Who am I to deny them? The decline of our imperial and colonial wealth means that we can no longer afford the national luxury of subsidising rural employment through enlightened social policy. National forestry policy has seen profound positive changes in the last 30 years. Amenity and conservation objectives do now have a seat at the top table, and the Forestry Commission has effective landscape design polices to promote broadleaf planting, amenity woodland and species diversity. Greystoke Forest, where Ali and I spent some of our happiest times hauling Sitka spruce, is now a good example of these enlightened policies promoting amenity, access, conservation, and diversity and I am grateful for these glimmers of hope. Back in 1978 only a few non-profit organisations maintained selective thinning regimes. The National Trust, the National Parks, the British Trust for Conservation Volunteers (now renamed TCV,) the Woodland Trust and a few progressive and well-heeled private estates agreed with me, and luckily there were just enough of them to keep us in business for a few years – until the price of timber dropped again, and hauling logs on piecework with horses, paid by the ton, became less of an ideological crusade or a forlorn gesture and more of an unrelenting struggle.

Until that happened we stayed optimistic and after a year on piecework at St. Croix and Greystoke we had established ourselves and paid off our debts. We pressed on, rolled the dice again and moved several rungs up the ladder. Thanks to the enlightened policies of the National Trust Forestry department, and their commitment to maintaining the deciduous woodland of the Lake District, we had the chance to live and work in the most magnificent wooded landscape in the country.

Three up on Ginger:
Brother Tom Lloyd and step-brothers Tom and Mike Barron.

11. Little Langdale - big timber

So, in the summer of 1978, mildly scarred but still full of faith and hope, we followed Ken Parker of the National Trust to look at some selective thinning at Parkamoor, a sloping woodland beside Coniston Water. The plantation comprised young mixed hardwoods, with larch and spruce planted as a 'nurse' crop. As we moved among the trees we disturbed a young roe deer, asleep in the grass about a yard away from us. It leapt up and disappeared in an instant, but the fact that there was any grass under the trees was a sign of our progress up the ladder - we would now be working in a more balanced environment than the sterile darkness of the Sitka spruce at Greystoke.

Ken wanted us to cut the spruce and larch and drag it down the hill to the roadside 200 yards below, without damaging the young broadleaf crop. The fast growing 'nurse' softwoods had served their purpose - they had grown tall and straight, forcing the oak, ash and beech to grow straight and tall with them so as to get at the light, leaving a stand of clean straight hardwoods when the 'nurse' crop was taken out. Ken would send in a chainsaw gang to fell the conifers, and they would convert my stacks of poles at the roadside. The extraction route was a long but gentle descent, so the going was easy, and best of all I would be paid an hourly rate! No more piecework for a while! We would start in a month, which gave me time to recover from the accident and move our whole rig to Coniston. Could he find somewhere for a caravan? 'No problem', he said, and took us to Boon Crag, the National Trust sawmill near Coniston, where we met the manager Mike Sim and the refreshingly lively foreman, Ken Ennion. We had a choice of site for our caravan – the sawmill yard, where there was a hard standing, running water, flush toilet, electricity, with shops and a pub nearby, or else a small clearing four miles away, in the middle of a wood in the middle of nowhere at High Park woods near Colwith Bridge on the back road to Little Langdale. No contest – the clearing had it, of course.

It was a truly beautiful spot, with a pair of Scots pine trees either side of a gateway which marked the entrance to the top of Atkinson's Coppice. The gateway looked South West towards the vast, brooding, rounded bulk of Wetherlam. Just inside the gateway was space for our caravan, a place to park the Morris Traveller and a place for the bender storage tent. Old stacks of thinnings provided firewood for the winter, and a few hundred yards into the wood was a trickling beck and a lonely tarn in a 'bee loud glade' - a clearing full of dragonflies. The young mixed woodland was full of open spaces and bogs, rich with sphagnum moss, sundews, bog asphodel and bog myrtle. Our annotated field guides to wild flowers, butterflies, birds, insects, and of course trees, were always to hand, and even 40 years down the line it is remarkable what a high proportion of our 'first sightings' of native species were made around that clearing in that summer. Half a mile away along the track was High Park Farm, we could get milk nearby at Colwith Bridge, with High Arnside Farm up the valley for eggs, and the metropolis of Ambleside only a few miles away. By a strange coincidence, our new home was a few hundred yards away from where, aged nine, I had come to camp with my family when my father ran his mares with a Fell Pony Society stallion on Oxen Fell in 1959. It was like the Promised Land. We shot back to Greystoke full of excitement, and started to pack up our gear.

It was hard to leave Joss, who had been so generous with his knowledge, but we had kept our side of the bargain as far as we could, and it was time to move on. The auction rooms in Penrith provided two vintage tin trunks, into which we packed all the tools and harness. We dismantled the bender tent and loaded the caravan so heavily that we wondered if we would ever get it out onto the road and if the tyres would stand the strain of the journey, and in mid-June 1978 we shook hands with Joss and were on the road again, bound for The Lake District.

A traveller friend Bobby Jeffryes, on his way back from Appleby Fair with a big horsebox wagon, loaded up Ginger, extracted our caravan from the forest ride and pulled our whole rig to Little Langdale. When we reached the clearing at Colwith, his truck was too big to get through the gate, so he backed the caravan part way in, and unhitched. We unloaded Ginger into his new field at Colwith Bridge in the valley below, which everything had everything we could wish for - secure fences, roadside access, plenty of grass, fresh water, and was surrounded by forestry which represented enough work for the foreseeable future. We paid Bobby for his time and his diesel, and waved him off, and walked back up through the woods to our new home in the High Park clearing.

High Park Woods, Skelwith Bridge. Ali at the window.

The caravan was still blocking the gateway, so our first job was to haul it up the slope and into position, but we had no vehicle to pull it. We had intended to wait for the National Trust tractor to do the job, but the tow-bar was sticking out into the road, so we could not wait for long, and in any case it had to be done by nightfall because our bed was invisible under a mound of harness and welly boots. Improvisation was called for. We used a dolly knot (see Appendix 4) and with two ropes, two standing trees and two dollies we were able to pull the 22ft trailer and contents, weighing about a ton and a half, up the slope and into place. It took us about two hours – unloading the caravan, planning, calculating, experimenting, tying and untying knots, pouring sweat, shovelling earth, chocking wheels, lashing spars, digging out stones, bending over, lying down, standing up, and easing our aching muscles, but we managed it. When the National Trust Landrover arrived, driven by our new friend the warden Peter Taylor, he refused to believe that we had done it by hand. There was no doubt that Ali and I made a good team, and that episode was another proof that two heads are better than one, and that willpower can achieve things that are obviously impossible – until you try, that is.

We lowered the caravan corner jacks, unpacked the tat, lit a camp fire, cooked our fish for supper, and pulled the mattress out of the caravan to sleep under the crescent moon, the summer stars and the crystal milky-way. Next morning we woke with the dawn chorus to see the great mass of Wetherlam to the west, glowing with reds and yellows, gloriously rising out the morning mist against the pale blue dissolve of a summer sunrise. As we lay under our blankets listening to the sound of unforgettable bird song, we watched the red dawn light slowly moving down the mountainside, as a pair of red squirrels came out and began running up and down the great Scots pine about five yards from our bed. Walt Disney could not have written that script, and even if he could, he could not have drawn it like it was, and anyway, he was not there to see it. We were there, and bliss it was in that dawn to be alive …

As it turned out, we did not start our work for the Trust at Parkamoor Wood, because the trees were not yet felled, but instead at Fletcher's Wood, near Colwith Bridge. It was 10 minutes from where we now lived, and five minutes' walk to Ginger's grazing field, which meant easier logistics.

It would be another couple of years before we settled into a house, so we were still running the whole operation out of the caravan and the back of the Morris Traveller. We would rise early, Ali would make the porridge and I would sharpen up and sort out the tools for the day. Through the back doors of the Traveller went the tin trunk of tools – chainsaw, fuel, oil, files, slasher, Yorkshire bill-hook, sharpening stone, timber tongs and turning bar. Then helmet, visor and ear muffs, safety gloves, leather working gloves, a leather apron (known as a *brat*), steel boots, first aid kit, measuring tape, spare shackles, neck collar, tethering chain, and spare hame straps. Last of all went the harness, the snigging chains, half a bale of hay, a bag of horse nuts and a feed bucket. And of course the all-important bait bag.

Ali, Ginger and Tom at Colwith

Horse logging is heavy work, and I learned from Geoffrey Morton that heavy work needs plenty of calories. Ali would bake three 1lb wholemeal loaves at a time in the tiny gas oven in the caravan, and each day I would make one loaf up into sandwiches – 1 cheese, 1 Marmite and

1 honey – with a bottle of water and a flask of coffee. (It took me about a year of regularly breaking my Thermos flask before we discovered a stainless steel version which could rattle around among the tools and survive. It cost almost a day's wages, but I still have it after nearly 40 years.) If I was lucky there would be Ali's cake or flapjack. This diet was my staple for about five years, and day after day I proved the old adage that it always tastes better when you are hungry. That bait tasted divine and I never got tired of it.

It was summer time, so Ginger was not interested in hay. The grass was so plentiful that he would often not want his morning feed-nuts either, but if he did want to eat I would sharpen up the saw and fettle up the harness while he chomped for half an hour. Gearing up the horse was a ritual. First I had to catch him, (I left his headcollar in place overnight) and lead him to the back of the Morris Traveller where I kept the gears. The main collar goes over the horse's head first, upside down, so the widest part goes over his eye sockets, and once over his ears it is swivelled round his neck so it is the right way up. Then the hames are fixed in place on the collar with hame straps. The rest of the trace gears go on with one swinging movement – the back-band and traces, hip straps, crupper, spreader bar and spreader straps would be gathered, swung up on to his back and then adjusted and fixed in place. The traces are hooked on to the hames at the front end and onto the spreader at the back end, with the snigging chain hooked onto the end of the traces. Then off to the woods.

I was in Fletcher's Wood for about two weeks. The slope was gentle, the poles were a handy size, not too big and not too small. The National Trust woodcutters had left a tidy job, and there was plenty of room for the stack at the roadside. I was keen to impress Ken Parker on my first job, so I rattled it out at a good speed, and by Friday when he came to see me, there was a respectable heap of timber. He was well pleased, and gave me the thumbs up, which was very fortunate, because the following week he brought a visitor – the National Trust Director-General, Jack Boles, the great chief from London. It was a surprise visit, and they were hoping to see the horse working, but unfortunately I was asleep under a tree when they arrived. Only mild embarrassment followed, and they could see that I had pulled out a good stack of wood and Ginger had obviously been sweating, but Ken never missed the chance to remind me about it.

Fletcher's Wood at Colwith confirmed what a champion horse I had in Ginger. It was our first job in steep and rocky terrain, and he never put a foot wrong. He would stand patiently while I gathered the load, then lean into the collar and quietly haul away, following the extraction route and working to voice commands. His greatest fear in the wood was the bot fly, or 'horse bee' as the older woodmen called it. The bot fly (sometimes confused with the warble fly) lays its eggs on the horse's front legs and when the horse rubs its nose on its legs or uses its teeth to get at an itch, the eggs are transferred to the mouth and from there to the intestines, where the larvae hatch, hook on, and begin to grow, causing colic and ulcers. A year later they pass out in the dung, then pupate for the cycle to begin again. My Victorian veterinary books suggested severe remedies for this parasite, and they can be a serious problem even in modern stables if not controlled. Horses may not be aware of the life cycle of the parasite but they hate them, and have evolved a very simple defence mechanism - they run whenever they hear one. In Fletcher's Wood we had our first brush with a bot fly, and I was unprepared.

Fletcher's Wood seemed to have a resident population of bot flies, which resemble a slender but noisy honey bee. Ginger heard this one approach before I did, and as soon as he heard it he rolled his eyes in alarm, and started to shift from foot to foot as if he wanted to run. By the time I saw the fly, it was too late - he ran. He had a load of poles behind him at the time, but he managed to move faster than the bot-fly and by the time he reached the roadside he had no load, and no harness except the collar, but he had escaped the fly and was standing quietly waiting for me to arrive. After that I was on the lookout, and I would usually manage to swat a fly before it landed and before Ginger took fright.

Apart from occasional bot flies, the 'clegs' or horse flies were a regular nuisance, with a nasty bite that came up in a lump on both horse and horseman. These long, brown, speckle-winged flies only came out in bright sunlight and the trick was to spot them as soon as they landed on an arm or leg, and kill them before they could bite. Midges were a constant irritation to me, but they never bothered Ginger, and wasps were more of a threat to my sandwiches than to me or the horse, although wild honey bees and hornets could bring the day's work to an end if we dragged a log through their nest. We could not blame them, but we would leave the wood at speed!

I discovered in Fletcher's Wood that Ginger was also alarmed and agitated by the unmistakable sound of the Maybug, also known as the cockchafer, a large and noisy forest-dwelling insect, with a voracious appetite for oak leaves, seedlings and roots, although harmless to horses. I discovered (probably from reading *Resurgence* or *Vole* magazine) that the Maybug was almost extinct after relentless extermination and the increased use of pesticides, so I had a dilemma. If I killed a Maybug to save Ginger a panic and possibly a nasty accident, could that be justified? What if I should kill the last Maybug in Cumbria - which was a real possibility - what then? Working alone for long hours in the wood gave me plenty of time for thinking and plenty to think about. Some days I pondered on all sorts of philosophical questions, and the horse bees and Maybugs got me thinking about the ethics of killing sentient creatures.

If the decision to kill such a creature was based on utility, convenience and expediency, it might make it easier to decide, but it still raised difficult questions. Killing a wasp which is about to sting is hardly a difficult decision, and very few people would tolerate a family of rats in the house, so speedy extermination would be the natural thing to do. A fox which destroys a dozen chickens for the pleasure of killing would be under a death sentence from many country people, who would sit up all night with a shotgun, although many people would take the view that the fox was innocent and doing what comes naturally, and consider the fox hunter to be cruel and bloodthirsty. Hardly anyone would want to shoot a deer that looked like Bambi, even though she had destroyed a thousand young trees and produced unacceptable amounts of methane ...

The prevailing attitude, at least among non-religious or carnivorous people, seems to be that killing pretty-looking creatures, especially furry ones with big eyes, is not acceptable, whereas killing ugly, smelly or stinging creatures is perfectly correct. That always seemed like pure prejudice or discrimination to me – quite species-ist, in fact, and to make it more complicated, these prejudices among humans change over time. The wolf was once feared, hated and almost exterminated but is now making a comeback in the affections of the public – at least among city dwellers who do not keep sheep. I never resolved the theoretical dilemma, but in practice I decided that I would happily kill a bot fly, but would probably not kill a Maybug, simply because they are a rare and threatened species, even if they made Ginger very nervous. Nothing is straightforward …

Either way, we were glad to leave the horse bees and the Maybugs in Fletcher's Wood, and when the job was done we moved half a mile up the road into Atkinson's Coppice, near Colwith Bridge, next to Ginger's field. Atkinson's was the biggest timber I had yet seen: Douglas fir and Western hemlock, second and third thinnings, but overgrown so that some of it was 60 feet tall, and I could not wrap my arms round the trunk. The terrain was steep, rocky, boggy and dangerous, with some windblow.

Although the woodcutters had done their best, this was the trickiest extraction I had yet attempted. Most of Greystoke Forest, where I had learned the ropes with Joss Rawlings, had been well drained and level or gently sloping, but Atkinson's Coppice was a mixture of crags, steep slopes, bogs and old ditches, all of which lay on the extraction route. It would have been impossible with a tractor, and uneconomic even with a horse on piecework, but the National Trust knew it was tricky, which is why it was overgrown. They had been waiting to find a horseman to get it done, and were happy to pay an hourly rate. I was happy to take it.

The steep slopes and rocky outcrops raised two problems – the sheer effort of getting a big horse up the slope between the trees, and the danger of 'over-run,' where the load would build up momentum and start to slide downhill faster than the horse. When this happened there was a danger of nasty injury if the poles struck the horse's legs, although Ginger had learned a clever trick from years of practice – if he felt the weight come off his collar while he was moving forward, or maybe if he could hear the load picking up speed behind him, he knew that the load was moving faster than he was, and he would step to one side.

Bill, Ginger and Tom Lloyd at Atkinson's Coppice

We had one or two narrow escapes, but no serious injury, and I learned all about what Joss Rawlings called the 'lazy man's load', which is not, as the name suggests, a light or small load, but one that is too big. It is so called because each trip to the bottom of the hill and the slow climb back up again might take half an hour, so there is always a temptation to load heavy. On piecework, an extra pole or two on each load might help the daily wage, but it is risky. A heavy load takes more effort to pull, and is harder to sort out if there is an over-run or a jam, so heavy loads could be a false economy. The temptation to overload was reduced by hourly pay, rather than piecework, but even on hourly pay the lazy man's load was always tempting because it meant one less slog back up the hill.

The Douglas fir and western hemlock in Atkinson's Coppice had a strong, pungent, characteristically sweet resin. The trees were covered in sticky bubbles, and oozed resin when the bark was scraped off during extraction. On rare hot days, after the trees had been felled and dressed out, the bubbles would burst and ooze what seemed like gallons of the stuff. I would come home at the end of each day with sticky, dark resin all over my hands, arms and clothes. The evening bath ritual required a fire in the clearing and an iron cauldron of hot water to get me clean. I seldom bothered to wash my work clothes, so had to struggle into my jeans on the morning after, when the resin had cooled and set hard, and my gloves would be rigid. The sweet smell that hung in the wood became nauseating, and although I would get used to it as the day went on, it would hit me every morning. Even 40 years later I cannot smell hemlock resin without thinking of broken harness, bruised arms, a weary back and jeans as stiff as boards.

In midsummer, we had a visit from my dad Walter. He turned up in his battered Ford Cortina pulling a small horsebox which carried Hades Hill Charlie, the Fell pony stallion that had pulled us round Yorkshire the previous year. We planned to use him in the wood, which suited Walter very well as it saved him work keeping the herd stallion away from his mares back home, and saved him the cost of feeding him all winter!

Unloading Hades Hill Charlie at High Park, Little Langdale

Walter was on his way to Torness Point, in East Lothian, where the Scottish Electricity Board had been given planning approval for a new Advanced Gas-Cooled (nuclear) Reactor. Torness had become a focus for environmental protest all over Scotland and the UK, and a Day of Action was called. At that time Walter sat on the National Festival Welfare Services committee and was not only experienced in First Aid, Communications and Emergency feeding, but was also a strong advocate for low impact alternative technology. He was on his way from Rossendale to Torness to observe the protest action, and we decided to take a few days off and go with him. There we joined 10,000 others who opposed the development of the AGR and intended to occupy the site and prevent the contractors from starting work.

The protest was a 'non-violent direct action', coordinated by SCRAM (Scottish Campaign to Resist the Atomic Menace) and there was much debate about how to get inside the fence without causing any criminal damage, which was an arrestable offence. A plan was devised to scale the fence by building steps using a wagon-load of straw bales, and about 1,500 people managed to get in by that method, but our carefully laid plan for peaceful protest was undermined just around the corner, where the more radical protesters just ripped the fence down and occupied the machinery compound. After prolonged debate, the radicals were eventually persuaded to leave by the moderates, and a kind of unity was preserved. Torness was an ideological melting pot and debating chamber, an alternative energy forum, a tribal meeting place, and a precursor of the Greenham Common protests and the Green Gatherings.

In early August we took a day off and climbed the Langdale Pikes. Sitting on the summit eating our sandwiches, we decided it was time for a holiday, and within a week we had packed up our rucksacks, loaded up our bikes and took the ferry to Belfast. We headed for Donegal, via the Antrim coast where were feted by the Loyalist shopkeepers, who had not seen an English tourist for a while because of The Troubles, and they loaded us up with groceries until our bikes could carry no more. In Derry we landed on the day of the Apprentice Boys March, where we skirted the gangs of teenagers throwing stones at the soldiers, and headed on out to the mountains. We found some mighty music sessions, got soaked to the skin on our bikes, and found a little bit of heaven when we camped on the beach at Clonmany, drank and sang with the Mayor of the town, and had our picture in the paper in the crowd near the finish line at the Clonmany Horse Races.

Back home in the caravan we found an extraordinary mailbag waiting for us. Offers of horse work had come in from all over the country - including Sheffield & Co. Timber Merchants and Geoffrey Morton in East Yorkshire. Ali had been offered work as a baker at the Kirkstone Gallery Cafe, about a mile down the hill at Skelwith Bridge, where she would spend her days making scones, shortbread and ginger cakes. I headed back to Morton's farm, to help to get the harvest. The National Trust had left a message promising what seemed like work for evermore, within half a mile of our camp. We bought a bottle of wine to celebrate our first full year of operation. We had no regrets, our cup was full, and it was all to play for.

High Park Farm, Little Langdale . (Photo: Tony Simpkins.)

Above Colwith Bridge – Footpath to our camp. (Picture D S Pugh)

12. Watto - the old school

He seemed to know everyone, and everyone seemed to know him, and they all knew him as Watto. John Watson lived in a bungalow near Skelwith Bridge, about half a mile from our clearing in the wood, with his wife Mary and his son Edward. He had been a horseman all his life, and though he once ran a small farm at Sunny Brow, near Outgate, most of his life he had worked snigging timber, so it was not long before our paths crossed. He had retired from farming to run a pony trekking operation for visitors, and to help him with his stable he entertained a company of young women who looked after the ponies in return for a ride out when things were quiet.

Watto was more or less alcoholic, and though he might go for weeks without getting drunk, he would be back on the whisky sooner or later. He carried a quarter bottle in his jacket pocket most of the time, and one wet day I sat with him in a pub while he drank 24 (small) bottles of barley wine in 12 hours. He was never once violent or aggressive when we knew him, although he had a cracking line in insults when needed. I never heard a bad word against him, but he conducted a kind of psychological war with Mary, his wife of 40 years. She was friendly enough towards us, although within a day or two she had warned us about his drinking and ordered us not to take him into a pub. Watto had been banned from driving for years, and so we and our Morris Traveller represented freedom and escape from Mary – into the woods, off to the Wigton horse sales or a farm sale, or just for a drive round.

Watto at Nibthwaite, measuring top diameters.

I spent most of September 1978 in East Yorkshire at Geoff Morton's, helping with the harvest and getting my first taste of work as a horse handler for films and TV, which later was to become my major source of income. Ali could not get time off from her job at the Kirkstone Gallery, so she stayed on at High Park, although she came to join me in Yorkshire for a few days. It was only when we returned to Cumbria that we became friends with Watto. Ginger had escaped from his field, wandered off, and

Watto had spotted him and brought him back. He was instantly friendly, and once he found out what we were up to trying to make a living with working horses, he adopted us. We soon got into the habit of calling in to see him whenever we passed by his house, and since he was disqualified from driving he would often suggest innocently that we take him off on a jaunt somewhere, and it was usually somewhere interesting. He found us work, he found us grazing, he found us harness and he introduced us to Westmorland farmers, woodmen, dealers, swill basket makers, coppice merchants, auctioneers, and horse people of all kinds. In particular he gave us the benefit of his knowledge of 'old school' horse management – veterinary tips, dentistry, and especially feeding.

Watto had his own system for feeding. I had learned from the Mortons that the best way to feed a heavy draught horse is a mix of rolled oats, hay, beans, and chaff. The whole mix was traditionally known as 'chop,' and although that would be a perfect diet for a horse on a farm that can grow all the ingredients, it was too impractical (and too expensive) for Westmorland hill farmers. Instead we followed Watto's way and fed sugar beet pulp, well soaked, with as much hay as the horses could eat. I liked to feed rolled oats and carrots by the sackful when the work was hard, which Watto considered an expensive luxury, but Ginger was never a 'good trougher' and it was always a challenge to get enough food into him to sustain the work-rate that we needed, so anything that he liked to eat, he got. We fed mineral supplements from time to time, in the form of a lick or a block, and in cold weather we even made up a hot bran mash with linseed oil, instructed by a treasured Victorian manual of horsemanship which we had found in a derelict house back in Hebden Bridge.

I collected old books on horse management in a casual way – we had little money to spare so I was never a serious collector, but I had a box full of battered and broken pages, mostly collected from junk shops, listing every ailment that a working horse might suffer, with some strange and wonderful remedies.

Fenugreek and elecampane seemed to be commonly used horse tonics, and they were even credited with magical curative powers, so I put them on a list for the next shopping trip to Kendal. The chemist knew nothing of elecampane, but I came home in triumph with a small bag of fenugreek seeds which I carefully mixed with Ginger's feed, only to find that after a tentative twitch of the nostrils he recoiled and would not touch it. I tried boiling it, drying it, grinding it and crushing it, but still he would not go anywhere near fenugreek. I found that elecampane ('Horse Meal' or 'Elfwort') grew on the roadsides, and we tried using the dried leaves and the ground-up roots, but never saw any obvious effects, although modern herbalists say that it is known to kill a range of bacteria.

Another veterinary curiosity in our ancient book was the 'medicine ball' or 'horse ball' - a mixture of whatever medicine was required, made up with flour or paste into a large pill, and administered with the help of a 'balling iron' - a curious metal gadget made of a 'U' shaped rod and either a ring or two straight bars, all attached to a wooden handle. The balling iron was put between the horse's teeth so that the medicine balls could be gently shot down his throat with a blow pipe made for the purpose. One of the old books even warned that the farrier should be careful that the horse did not blow back!

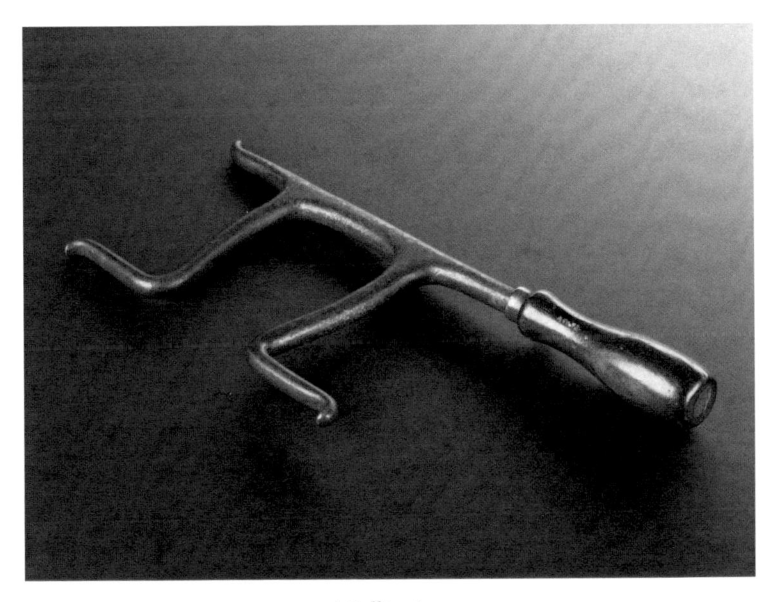

A Balling Iron

Another type of balling iron was used to hold the mouth open when filing teeth, and it was Watto who first instructed us about horse dentistry. When Ginger suddenly stopped eating one morning in the wood, Watto quickly diagnosed a sharp edge on a back molar tooth which had raised a painful sore in his mouth so that he could not bite or chew. We drove back to Watto's stables where he produced a bundle of tools wrapped in an old tea-towel. Among them was a set of horse tooth files, and after 20 minutes of struggle to hold Ginger's head still, the offending edge was rasped away and a day later he was eating as normal.

Watto showed us how to tell the age of the horse from its teeth, and told us all about 'wolf' teeth – small peg-like teeth, usually in the upper jaw, but which can cause problems when they grow in the lower jaw. He told us about a poor-looking horse which appeared in the auction, listless and thin as a rake. He spotted that it had a wolf tooth, bought the horse for a song, and then extracted the tooth as soon as he got it home. The horse started eating again, and went on to become a champion. (We later heard the same story in a Gypsy folk song, which was probably where Watto heard it in the first place.) Horses don't like dentistry any more than humans, so tooth care is often overlooked, but is an essential chore.

Watto's old tea-towel also revealed a fleam. Most of the old horse books had pictures of a fleam – a surgical device used for opening a vein so that blood could be taken. The practice of 'bleeding' has largely died out, although 'blood doping' is occasionally discovered in racehorses even now. The fleam is a razor sharp triangular blade on the end of a shaft. The blade is laid along a vein (not across it) and then struck with a small mallet made for the purpose. The shape and sharpness of the blade could penetrate the thick skin of the horse and make a neat incision, which could be easily stitched up after the blood had been taken. Bleeding has been used for horses, and humans, for thousands of years, and although opinions have changed as to the medicinal value, it was once a common treatment, and taking a few quarts of blood with a fleam was the standard veterinary remedy for laminitis, fever and various swellings. Watto's fleam, which had belonged to his father, was a beauty – three bright steel blades set like a pen-knife in a horn mounted handle.

A Fleam

I never had any wish to try bleeding a horse, but I was interested in the whole technical and mythological pagan mystique around the 'horseman's word' (also known as 'horse whispering') and the 'drawing oils' that could supposedly quieten the wildest of horses. The 'Frog's Bone' for example was reputed to have magical powers. According to our guru and favourite Victorian writer, George Ewart Evans, writing about traditional practices in Suffolk, the aspiring horseman would go out on the night of the full moon and search for a frog. When he found it he would kill it and leave the body on an anthill for the flesh to be removed. When that was done he separated the pelvic bone from the skeleton, and kept it in a leather pouch full of herbs. When approaching a difficult horse, the bone would be hidden in the hand, then shown to the horse, if necessary rubbing it along his body. The horse would then fall under the horseman's spell and become docile and compliant. Hmm ... I never tried that one either. Maybe ... and maybe not.

Although I occasionally asked very diplomatically about the 'horseman's word', the 'Frog's Bone' and the rest, none of the Northern horsemen I came across knew anything about a secret 'Society of

Horsemen' or magic tricks with herbs, although of course if they did belong to a secret society they would not have told me anyway. I heard of occasional funny handshakes, and like Watto, all the older horsemen I knew had a few tricks of the trade, but I concluded that the 'Society of Horsemen' and the 'secret mysteries' were probably South-country ploys to fool the unwary, the folk-lore collector or maybe the boss. (I remain to be convinced, but I am open to persuasion. George Ewart Evans makes an almost convincing case.)

The only items used with any regularity in Watto's mysterious tea-towel full of implements were his gelding irons and his twitch. The twitch was a loop of strong cord on the end of a wooden handle. It was used to quieten a difficult horse for surgical or veterinary procedures, or if it would not stand for shoeing for example, but not by any mysterious or secret hex – the twitch used practical mechanical rather than magical means. Once the wild horse was haltered and more or less standing still, the loop would be placed over its sensitive upper lip and twisted very tight with the wooden shaft. Most horses will stand very still when this is going on. Watto assured me that the pain in the lip distracted the horse, although modern veterinary practice suggests that the effect is caused by the intense pain releasing endorphins, which have a calming effect.

I cannot image there were many endorphins released by the gelding irons, which were a pair of steel clamps used to close the blood vessels before castration of a colt. With blood vessels clamped, the testicles were removed with a razor sharp knife, and a pair of red hot wedge-shaped irons bars on long wooden handles were used to cauterize the wound.

Unlike the fleam and the Frog's Bone, both the twitch and the gelding irons are still in use today by vets, although welfare legislation clearly prohibits the infliction of any unnecessary pain and suffering. Castration was, and remains, a regular and quite normal procedure in the horse world, and a skilled man could carry out the operation in under a minute, with no blood and no obvious pain, so that the new gelding would be quietly grazing five minutes later. Geldings have almost all the power and stamina of a stallion, without the troublesome hormones that give them both the incentive and the ability to jump fences, break gates, bite, kick, and fight each other so as to see off any rivals for any nearby mare that is in season, or 'horsing' as the old boys called it.

More important to me than the Frog's Bone or the twitch for getting my horses to cooperate was finding good hay, which became a minor obsession. The best hay at Geoff Morton's was known as seed hay, hard hay, or just 'seeds' and it was a species-rich grass mixture with high protein content. It was also relatively hard in texture - one reason why it was chopped up and mixed with the corn and chaff to make 'chop.' Most of the hay in the central Lake District was meadow hay - softer in texture, more leafy, and more suitable for sheep. By the mid-1970s, wrapped silage and 'haylage' was becoming more common, particularly in the Westmorland climate of wet summers and a short growing season. As big bales of wrapped silage became the norm, so finding good hay in small square bales became a priority. We knew a few farmers who made their own hay, and one or two who brought wagon loads in from Yorkshire, and in times of shortage I could always buy from Jordans, the feed merchant in Windermere. They always kept good quality hay although the price was high. Since we had no hay barn while living in the caravan, Watto and I would make a regular weekly trip to the merchant in Windermere or up into Langdale, or anywhere we could find good hay. We would put the back seats of the Morris Traveller flat to the floor and stuff the body with bales so that the back doors would not close, then lash a few more to the roof-rack and trundle home along the lanes, to store them in the wood on pallets under tarpaulins.

Ali and I became best of friends with Watto. We had learned about driving horses on the road from my dad Walter, I learned about working heavy horses on the land from the Mortons, I learned snigging from Joss Rawlings and we learned about forgotten traditional techniques from books, but Watto taught me most about life, about laughter, about 'the crack', having fun, about old Westmorland ways, and the pleasures that money can't buy. He was our 'main man' and I would go to pick him up for the day whenever I could. He was just good company.

Through the winter of 1978/79 Watto was busy looking after the dozen ponies he used for his trekking business, breaking in young stock, and trying to get his son Edward into some kind of work, so we did not see each other every day. Later, when we had settled into a cottage in Sawrey and he had given up his ponies, he offered to come and work for us for nothing, just because he enjoyed the crack. I offered to pay him but he was reluctant to take any payment as he knew that money was tight, but I insisted and persuaded him to take a token amount for a day's work. I paid for his tobacco and his dinner, and put a ten pound note in his pocket as often as I could, so he preferred to spend his time in the woods instead of 'sitting at home eating 'taties' as he put it.

The extra set of hands made a big difference, and Watto was such good company I missed him when he wasn't there. Apart from knowing the horse job so well, and teaching me some good old tricks, he had an unending repertoire of stories, mostly from first hand experiences, and he almost never repeated himself. In his late teens he had worked as a stallion walker, leading a stallion from one farm to the next to cover the farm mares and get foals for the next spring. Every young farm man wanted that job - looking after a big fine stallion horse, escaping from the conventions of home and lodging in the farmhouses along the way, enjoying the crack with the farmers, and always the centre of attention for the girls, for whom the good looking young buck and his virile horse were (he said) irresistible. He described that time as the best years of his life, and who would argue? He told me a tale that when he married Mary, a Langdale girl, a Langdale farmer had come up to him at the wedding and observed that it was unusual for him, an off-comer, to marry into Langdale, which was still a bit of a closed community. "Aye" said Watto "she warned me that there was nobbut two ways into Langdale, you had either to fight your way in or f*** your way in, and by God I had to do ba'ith."

In his youth he worked as a horseman for Mrs. Heelis, better known as Beatrix Potter. Watto was fond of a scurrilous tale, and one day while sitting in the car in the rain he told me that in her later years Mrs. Heelis would dress roughly, and looked like a shepherd. Sometimes on a cold day she would grab whatever headgear came to hand and she appeared once wearing a tea-cosy on her head. She had a few warts and a certain amount of facial hair, and the village children put it about that she must be a witch. One morning Watto appeared in the yard at Hill Top Farm with his stallion horse, come to do its job, and Mrs. Heelis was standing in the doorway giving the orders for the day. Leaning up against the wall by the door was a birch besom broom. "I see you've got a new bike" he said. She was not amused.

It was thanks in part to John Watson that, years later, when Beatrix Potter's farm manager Geoff Storey, sold up at Hill Top, we went with Watto and my dad Walter to the farm sale. Walter could not resist buying her gelding irons, which he brought home in triumph. He paid very little – probably because no-one but a horseman would know what they were, but Beatrix Potter's gelding irons now hang by our fireplace as a reminder of my days with horses and my old marra, Watto.

Watto had worked as an axe man in his youth, and during one of our many fireside sessions, he told me that in his opinion, felling trees with an axe was the hardest work a man could do, and that axe men seldom worked at that trade beyond the age of 35. He also told me about the games of strength and courage played in their breaks. One game was to raise a felling axe, one handed, at arms length, and to bring it down so as to split a matchstick, set on end in a tree stump. Bets were placed, and the first man to do it took the kitty. He described how they would find a tree which was standing alone so that the branches reached all the way down to the ground, and scramble up the trunk. When they got near to the top they would clamber onto the outside and slide back down to the ground on the branches.

One day, while sitting in the car waiting for the rain to stop, I asked Watto what he would choose to eat if he could order any meal he liked. He thought for a short while and said 'Well, you know Bill, it takes a lot to beat a dish of stewed cheese.' I asked his wife Mary how she made it, and it could not be simpler: put some milk in a dish, add some cheese, and put it in the oven. Brilliant!

Mary had a hard time with Watto, no doubt. They lived as simple as could be – a paraffin heater to heat the house, with maybe a bag of coal at Christmas, minimal furniture, the same clothes day-in day-out, porridge for breakfast, sandwiches for lunch, lamb or pork with cabbage and potatoes for supper, roast beef on Sundays, no takeaways, no dining out, and no holidays.

Watto's fondness for the bottle got him into plenty of trouble, and I heard the story that one day after a great row at Sunny Brow he left the house and drove down the nearby Outgate Inn. When he did not come home, Mary strapped a sledgehammer to the crossbar of her bike, cycled down to the Outgate and wrote off his car with the sledgehammer to stop him from driving home. It was not just out of spite, but to stop him killing somebody by driving home drunk. She had a reputation as a dragon, but with us she was friendly enough – we took Watto off her hands, kept him out of the pub (mostly) and brought him home cheerful.

In the autumn of 1978 we started work in the steep woods at Yew Tree Tarn, and not long after we started there came unusually heavy snow. On some days we were unable to travel about the lanes to get to work, or to feed Ginger and the Fell pony stallion, Charlie, who were grazing in the pasture beside Yew Tree Tarn. Watto suggested that we asked Johnny Birkett of nearby Yew Tree Farm if we could move our caravan onto his ground while we were working there. Watto came with us to make the introduction, and Johnny immediately offered us the holiday cottage adjoining his farmhouse, which he had closed up for winter.

Johnny and his brother Danny farmed next door to each other, and they were a kind, generous and helpful family. At that time we had no horsebox, but Danny would lend us his cattle trailer without a second thought, and he would even tow us about if he could. Maybe they felt sorry for us, living in the caravan in winter, but Watto's introduction was like a passport into the close-knit Westmorland culture, in which neighbours helped each other on principle, so they offered the cottage at a low rent, and we moved in to Yew Tree Farm Cottage before Christmas.

The Birketts were hospitable, and the cottage was convenient for feeding horses and for getting to work, but we had in fact been more comfortable in the caravan, which could be brought up to sauna temperature in about half an hour on a few logs, whereas the cottage took all day to warm up with an electric heater, so we struggled to dry our clothes. We were glad to be only a few hundred yards away from the horses, but as soon as we finished the job at Yew Tree Tarn, Ali and I planned to move on again – to Penny Rock Woods, near Grasmere.

Oak Butts at Penny Rock Wood, Grasmere

13. Grasmere - *dynamics*

The job at Yew Tree Tarn was right beside the main road from Coniston to Ambleside, and we often had an audience of visitors who stopped to watch the spectacle. One day the Lake District National Park head forester, Jon Williams, paid a visit and introduced himself. He wanted to talk about horse extraction, and of course we obliged. (Another advantage of the hourly rate over piecework is time to chat ...). We spoke the same language - Jon already knew all about the advantages of using horses in the wood - low capital cost, low impact, access to difficult terrain, and less damage to the standing crop. He had a job for us - thinning the oak wood at Penny Rock near Grasmere. Would we take a look at it? (Does the Pope wear a funny hat?)

Penny Rock Wood moved us up a league. This was bigger and much heavier timber - oak sawlogs, very different from the conifer thinnings we had been working up until then. The trees had been felled and dressed, and most of the bigger branch wood had already gone as firewood, so it was a straight hauling job, paid by the hour, in a high quality amenity woodland. The contract had first been taken by a timber merchant with a D2 Caterpillar tractor, but for reasons that were never clear, he disappeared and we were brought in to finish the job.

The first difficulty at Penny Rock Wood was commuting from Yew Tree Farm Cottage to Grasmere in deep snow. Travelling a few miles down the hill to Skelwith Bridge and then a few miles more up over Red Bank may not seem like much, but when we had to keep the horses fed twice a day, seven days a week, in winter conditions, it was not ideal. Red Bank was steep and icy, impassable in freezing weather, and the 'long way round' via Ambleside was usually heavily congested and busy.

We asked around Grasmere village trying to find lodgings, and in the Post Office, thanks to the lady behind the counter, the village grapevine worked wonders and we were told about Aya, a mystic and astrologer who had been a co-founder of Harvest, the first vegetarian restaurant in Ambleside. She had sold her interest in the restaurant and had purchased Dry Close, a fine mansion house above Dove Cottage, a mile out of Grasmere on the way to Penny Rock, where she planned to open a spiritual centre. She had re-named the house *Ishvara* and painted a huge pentagram on the floor of the library. She might have rooms to let.

Ten minutes after we heard that, we were knocking on her door. She thought we had been sent by God. She had been troubled by some kind of feuding in the village, and that very day someone had left a dead rabbit on her door step. She took this as a serious ill-omen, and had prayed for an hour for someone to come to her rescue, to stay in the house with her and protect her from dark forces. Half an hour after that we knocked on her door looking for lodgings, and her prayer was answered. We were almost local, and we were practical, helpful, friendly, and with a colourful history in theatre, poetry and horses, I was a lumberjack horseman, complete with beard and checked shirt, Ali knew about herbs, baking, gardening and chickens, and neither of us was frightened of dead rabbits. (We assumed that the cat had done it.)

Aya asked us to move in that very night. We drove back to Yew Tree Farm Cottage, feeling yet again the blessing of good fortune, packed up the Morris Traveller, and shifted to *Ishvara*. We moved in, stoked up the Aga until it was glowing red hot, hung our wet clothes in the drying room, took a long hot bath in a bathroom the size of a bedroom, and slept in a huge brass bedstead in a fine room on the west wing overlooking the glories of Grasmere. We had breakfast in the dining room and thought we had landed on our feet yet again, (thank you, Goethe) only to be relegated to a freezing attic after our first night of splendour. So it goes.

Next day we walked Ginger and Hades Hill Charlie over the hill from Yew Tree Tarn, through the misty valley, up to Loughrigg Tarn, down Red Bank, over the wooden footbridge where the river Rothay runs out of Grasmere, and into Penny Rock Wood, where we turned them into the huge field by the car park. Charlie the stallion had been grazing at High Park, and then with Ginger at Yew Tree Tarn, and now we planned to work him at Penny Rock. We found a collar to fit, borrowed John Watson's lightweight snigging gears and Ali joined me in the wood, getting the smaller branch wood with Charlie while Ginger pulled the big butts.

Ginger, pulling uphill at Penny Rock Wood

The terrain was mixed - rocky, boggy and uneven, and for the horse-logger, terrain is everything. Once you have a good horse and a good set of gear, terrain is all that stands between you and making a living. Although the horseman usually gets the roughest terrain, he only gets it because the tractor men can't or won't tackle it - if it was easy work the machines would have it. On level terrain a good strong horse is less productive than a tractor, but the lower capital cost and operating cost of a horse helps to make up some of the difference in productivity. On steep terrain, for smaller timber which does not justify rigging a winch, the horse is more productive. In any terrain it is hard graft and it is only a strong preference for working with horses that keeps the horseman afloat.

It is easy to be led astray in discussions about the relative power of the horse, due to the many variables which influence the dynamics, and most of those variables concern traction and four different types of friction. The most significant variables are: two wheels or four wheels, or no wheels at all, the diameter of the wheels, the weight of the vehicle, the ground surface, the type, breed, age, stamina and experience of the horse, the type of shoes on the horse, the steepness of the gradient, the length of the pull, the required acceleration, the skill of the driver, the length of the rest periods, the feeding regime and even the weather. Clearly 1 ton on wheels is very different from 1 ton being dragged along the ground, which is different again from a 1 ton weight in a measured lift or 1 ton of Newtonian resistance in experimental conditions. The greater the friction from dragging a log along the ground, the greater the traction needed to move it, so using a wheeled vehicle or a sledge, and even changing the type of lubrication on the axle can be shown to be significant variables. The calculation is further complicated by exaggerated claims by some horse owners, breeders and loggers who are keen to impress, and by the fact that many written sources do not bother to define the variables, or they simply repeat wrong information found in reference books.

What follows is an outline of the subject, and is not intended to be an exhaustive or definitive explanation. My purpose is to show aspiring horse loggers that the only way to find out how much your horse can pull is to try it, measure it, and then avoid any guesswork. If you have a feel for physics, you can do calculations using various mathematical formulae such as the 'coefficient of kinetic friction' but if not, you can use a simple formula which I developed for myself: 'Work for a week, and then check the weigh bridge ticket. Make a note of that number, and believe it .' You might not like it, but any other number is certainly wrong.

The term 'horsepower' is widely used to describe and measure traction. James Watt, pioneer of steam engines, calculated that 'one horsepower' is equal to about 750 units, which we now call watts. In fact Watt's calculation relied on an 'estimate' of the friction in his steam engine, and modern experiments show that his engine could in fact only produce two thirds of the claimed 'one horsepower'. (See the glossary for full definition of 'Horsepower.') I have never tried to repeat Watt's experiment in a laboratory, but I always suspected that the popular conception of the work that can be done by 'one horsepower' is wide of the mark. I know that in a straight pulling contest, my one horse Ginger could walk away with a 3 horsepower lawnmower and not stop pulling until bed-time.

All this shows is that the real world of horse logging is very different from laboratory conditions, and that the one cannot always be used to judge the other. Laboratory figures with weights, scales and calculators may be accurate, but they certainly do not take account of the various inefficiencies which characterise logging - the steepness of slope, the wetness of bogs, the height of the brash, different types of harness, snags and jams, feeding, and even the species of pole, all of which affect the amount that can be extracted in a fixed time. As a final complication, it has been shown that the maximum horsepower developed by a draught horse in pulling contests is determined as much by the speed of the pull as by the weight of the load, so using 'one horsepower' as a standard unit of traction is full of snags, jags, pitfalls, bogs, diversions, and difficulties - just like horse logging in fact.[1]

Horse pulling contests can provide interesting information. Horses working in teams are less efficient than horses working singly. This loss

---

1 *Animal traction: guidelines for utilization. Michael R.Goe and Robert E. McDowell 1980*

ranges from 7.5% for a pair to a loss of 37% for a team of six. The advantage of working a team comes from the fact that in a team, less human power is needed to gain more traction, since one man can work six horses. A pair of horses can develop between 14 and 30 horsepower over a ten-second pull, and in one contest a single horse pulled 900 kilos over 2000 metres in 5½ minutes, at a trot! It has also been shown experimentally and fairly conclusively that a smaller horse can pull a higher percentage of its own body weight than a larger horse, although this only applies at extremes of performance. Oxen can pull more than horses, and are still used for logging in some parts of the world.

Meanwhile, back on the ground, as a 'rule of thumb 'it is generally reckoned that a horse can safely pull its own weight for short distances on wheels in rough terrain, i.e. bad roads, sandy soil or level fields. On a hard flat road it can safely pull three times its own weight on wheels, although even this can be exceeded in favourable conditions. Stamina is important – some horses can pull heavier weights for short distances, but are not so good at sustained road work. A horse can produce a certain fixed amount of applied energy each day, and this can be in short or long bursts. A typical workhorse can pull ⅛ its weight at 2 mph for 10 hours, or at 4 mph for 5 hours, (or ¼ its weight at 2 mph for 5 hours, or all of its weight at 2 mph for 1¼ hours, etc.)

The best information I found on British forestry horses came from the *Forestry Commission Census of Harvesting Methods, 1967.* In that year, in Scotland, thirteen horses pulled out 19,000 tons of wood. Assuming that the horses worked 260 days per year (which allows the horse no holidays, just weekends and a few days off) that averages out at 5.6 tons per day per horse. If 20% of the time is deducted for bad weather, the average would have to rise to 7 tons per day, or about 1 ton per hour of actual work, which accords with my estimate based on practical experience. In good going I might aim to get 200 cubic feet (about 7 tons) on a good long day in good terrain with no breakages. In reality I took longer breaks, and a fair amount of days off to do other things, and the terrain was always difficult, so my yearly total would be well below the Forestry Commission 'staff' horsemen, who worked regular hours with very little time off.

One thing is certain from my experience: after taking into account the size of the horse, and 'all other things being equal' (what a great get-out that is!) it is the terrain that determines the average weight of the load your horse can pull. On gentle downhill slopes, with a heavy horse and with no specialised skidding equipment and no obstructions, I could get 6 cwt per load or 2-3 cwt with a pony. This average would drop sharply on an uphill pull, but on a steeper downhill slope, with no stocks, no bogs, no rocks, no water crossings or ditches, and no uphill sections, a man and a big horse can rattle it out, especially pulling big single butts that are quick to hitch up and have been well dressed out. 10 cwt would be a good load. The largest load I pulled at Penny Rock was a 28-foot oak butt with a diameter of 20 inches, weighing just under a ton, but this was exceptional – the butt was well placed on a downward slope, the pull was started on a roll, and we had to travel only 20 yards to the roadside.

On uphill slopes with a good track, the efficiency can be increased by lowering the ground friction through the use of specialised equipment – skids, sledges and arches, but these tend to be expensive and unstable on rough terrain. I tried out a Norwegian skidding arch at Greystoke, and found that unless they are used on a long haul on a good track, the arches were generally more trouble than they were worth – they are expensive to buy, and clumsy and awkward to use on the rough and steep terrain. If the woodland owner buys the equipment, and pays by the hour and not by the ton, a skidding arch could make the work a lot easier, on the man and on the horse, but otherwise I found it to be an extravagant luxury. The only exception was the Canadian sledge bolster which I made myself, which cost me very little and was small and light enough to drag back uphill and lift into position next to the load.

The slope is not the only factor when considering terrain. Bogs can be dangerous, but luckily the horse usually knows best – he knows when to keep well out of a bog, even though it might look innocuous to the human eye. You could try to force the horse to go across a bog, but that might be a bad mistake. If the horse is wary, think very carefully before you cross, especially when loaded! It is not just the danger of the horse getting stuck – which can be fatal – it is the risk of his pulling a muscle, twisting a knee or straining the tendons when trying to get out. If a horse panics when loaded, damage will be done somewhere.

After checking for bogs and ditches, I would sometimes spend hours at the start of a job clearing a good track out of a wood so as to avoid obstacles – cutting stocks down to size, cutting out overhanging branches, clearing brambles or taking out whole trees to give a good line without too many turns. High stocks and sharp rocks are a dangerous nuisance, and brambles can be surprisingly troublesome. After you have once tried to extract a load of small poles that have snarled up in a bramble bush you don't want to do it again.

The fewer bends in the route and the straighter the pull the better, especially on steep ground, so as to avoid the 'fish-tail', which happens when a long pole pivots around a tree or a rock. A pivot can be very useful to get a clean change of direction, but once the leading butt has cleared the pivot point by a few feet and the horse changes direction, the tip of the pole can be 15 to 20 feet away and can travel in a great arc at tremendous speed. If you are standing in the way it will probably break your legs in passing. Natural forces are merciless. Even a light knock from a fast moving tip can damage your shins fairly badly, and I have the scars to prove it. A few sentences out of a book can warn you, but only the experienced eye can avoid it happening in real life. Luckily it is usually a fast learning curve, and tough shin pads became an essential piece of my woodland kit.

The steeper the slope the better, up to the point where it is so steep that the pole will over-run and start to move downhill faster than the horse. If this happens it will usually overtake the horse, and pull his back end round, so he is then facing uphill, or at least sideways. This is not a good idea, but unless it is moving very fast, the weight of the horse will stop the pole and he won't be dragged downhill. Sometimes the pole will unhook itself as it passes and keep going for a short distance. More often the horse will pass to one side of a standing tree and the over-running pole will pass to the other. This has the advantage that the standing tree and not the horse takes the weight of the runaway, which saves any injury. But the pole has then gone the wrong side of the tree, and usually there is no way out by that route. When this happens the pole has to be cut free, but the chainsaw is usually at the bottom of the hill by the stack, or else at the top of the wood where you have been felling, or it is in the van, or out of fuel, or it is broken; so you have to go and fetch it, and/or fix it. Sometimes the horse cannot move backwards up the hill to unhook the chain until you have either cut the pole (or poles) or pulled the load back using brute force. If the saw is not available you must use the bill-hook and hack it out, but one way or another you must cut the pole, free the horse, re-fix the snigging chain, cut a new route, and start again. Over-runs are exhausting. Over-runs on steep slopes with big timber can be fatal.

This may all sound discouraging, and suggests that horse loggers need a cloak of invincibility and saintly patience before they start. I had neither, so I learned to be careful, although patience was harder to learn. After a while I came to expect snags and frustrations, but I never took them easily. As each progressively more strenuous attempt to overcome a snagging problem by muscle-power would fail, I would become progressively weaker, and the chances of success would become slimmer. I would feel the pressure rising and occasionally (probably about once a month at the full moon) I would eventually lose my cool, throw down the tools, rip off my shirt, batter my helmet on the rocks, and let out a strangled yelling growling scream of frustration. Not pretty at all. I tried not to do it if there was anyone nearby, but now and again I just lost it, and did it anyway. 'Childish' you might say – 'grow up!' 'A long way from George Sturt and George Ewart Evans ' you might say. Exactly right. That is one reason that most woodmen prefer tractors.

The steepness of the slope helps with extraction, but the steeper it gets, the harder the work to get back up the hill, especially as the distance from the production bay increases. My horse Ginger, at slightly over 16 hands, was a little too big for the really steep slopes in the Lake District, and he had to stop for rests often, but then so did I. Little Charlie was nippier, but could not pull such big weights and he had to make more trips, so it evened out in the end. In general the ideal size for a snigging horse in the steep Lake District woods would be about 14.2 to 15.2 hands.

Once I had climbed up high into a steep wood, I developed a habit of refusing ever to set off downhill unless we were loaded. So if for example I had worked my way up a slope and gone past the lowest point of the 'next' pole to drag out, I would never take the horse back down again unloaded even for a few yards to get that pole, but would instead go further up and get one that was higher. Whether it made any difference to my efficiency in the long run I never knew, and Watto found it a bit eccentric, but that was my way. Getting myself and a big horse up a steep slope thirty times a day, then lifting and heaving the poles on the way out and then again at the loading bay did keep me fit, but it put me off hill-walking for life.

When working in difficult terrain, the welfare of the horses has to be a higher consideration than the thickness of the pay packet. Cliffs, bogs and rocks are obvious dangers, but sharp spikes, deep drainage ditches, loose loads, broken harness, and lost horseshoes can all happen without warning, and cause nasty injuries if not spotted in time. At Penny Rock the terrain was mixed. It was not steep, but it was boggy in places and the work was hard on the horses due to the size of the wood. We were pulling big oak butts weighing over half a ton, which would have been no problem on a downward slope, but on level ground or uphill they were heavy work. The initial pull is the hardest, and once the log is moving it takes less effort to keep it moving, so I would usually wrap the chain once or twice around the log before hooking on, so that the inertia was overcome with a twist and a roll, rather than a straight pull.

Even so, after a few days I had to call Jon Williams and ask if I could cut some of the big butts into shorter sawlogs. It lowered the value to him, and it damaged my pride, but it saved damaging the horse and it allowed me to finish the job. Jon agreed without a murmur, probably because Penny Rock was an amenity wood and the avoidance of machine damage was worth more to him than the lost value of the full-length sawlogs.

Although the work was heavy, the pay was good, our lodgings at *Ishvara* were excellent, Aya was helpful and friendly, and the horse field was nearby, secure and rent-free, but the lack of a proper stable was about to become a serious problem. In January the field did not catch any sun all, so it was always cold and frosty, and we would work short days so that Ginger could cool down before nightfall. We had been there a month when I noticed a small patch of hair coming off Ginger's flank. Looking up in my veterinary books, it looked like some kind of mange, so I called the vet. He sent off a sample for analysis, and next thing we knew there was a policeman knocking at the door.

Bill and Ginger, Ali and Charlie, with oak butts. Penny Rock, Grasmere, 1979

The veterinary laboratory had diagnosed Parasitic Mange, a statutory 'notifiable disease' since 1911. Even though not a single case had been reported in the UK since 1962, and although the disease was officially 'eradicated' in 1951, the laboratory was adamant they had not made a mistake. The animal health department at DEFRA (or MAFF as it was called then) must have got very excited as they dusted off the rule book.

And what a rule book it was. First we were served with an immediate 'restriction notice', and then a formal 'quarantine notice', forbidding any livestock movement on or off the field at Penny Rock. Next we were presented with a sheaf of forms, which required our personal ID details, employment status, Ginger's details (breed, age, height, colour, markings, etc) and the dates and times of all our movements for the last month. Then the Ministry vet arrived. The rules stated that affected animals had to be isolated and treated, and the detention notice could only be removed *"when the inspector was satisfied that all horses, asses and mules on the infected premises were free from parasitic mange and when the premises had been satisfactorily cleansed and disinfected."*

We were required to purchase a bottle of (expensive) medication - presumably an insecticide - with instructions to apply it to Ginger as a whole body wash twice a week. On welfare grounds alone, this was ridiculous. He was in an open field with no shelter, and no hot water. The sun never rose above the horizon of Red Bank all day, so the temperature was constantly below freezing for weeks on end. I was not allowed to move him to a stable, or even move him 200 yards into the next field where at least there was some sunlight (pale and wintry though it was) because a public footpath passed nearby, and the rules did not allow him to approach within 50 yards of a public footpath.

Anxiety took hold. I informed the Ministry that I refused to comply, on welfare grounds, and challenged the diagnosis. For a while it looked as if Ginger would have to be destroyed unless we agreed to scrub him down with ice-cold water in a freezing field twice a week. My patience with the bureaucrats was wearing thin, but the regulations could not be broken and we were running out of money because we could not work.

At this point I nearly quit the whole thing - we were living hand to mouth, so we had no income unless we were working, no savings, no land, no stable, no money to pay vets, no lawyers to make our case in court, and unable to afford to erect a temporary shelter where we could house Ginger while his 'deadly' (i.e. harmless) hair-loss was treated.

In the end, the power of the state could not be denied, and we had no choice but to comply. We carried buckets of hot water about half a mile, washed him down, then dried him off as best we could with sackfuls of old towels and fed him as much hot mash as we could get in to him. He survived, and so did we, although neither Ali nor I can remember quite how.

Looking back it seems trivial, but at the time it was a big blow. For the first time we were paying an unreasonable price for being out of step, out of our rightful time. We seemed to be fighting enormous odds, just to do something which seemed to us so sensible, natural, simple and rational. All we wanted to do was to work horses in the woods, and suddenly we were a public enemy and had policemen checking our movements, all because of a bald patch on a horse, about 2 inches in diameter. It was insane and illogical. Six weeks was regarded as the maximum time the offending parasitic mite could live without a host, and no case of parasitic mange had occurred for 17 years, so it defied common sense as well as logic, but the Ministry were having none of it, and the vet muttered about 'mutations' and 'resistance.' It was my first taste of the bureaucracy I was to meet later during the foot and mouth epidemic of 2001, when the DEFRA officers and vets were always charming, friendly, meticulous, and often hopelessly irrational. I spent a whole morning composing my best wail of complaint to the Ministry of Agriculture, mixing invective with appeals to their common humanity, and pointing out the anomalies in the rule book, but I had no reply. The Parasitic Mange Order of 1911 was revoked just a few years later in 1983, although I doubt that my letter made the slightest difference.[2]

Eventually the crisis subsided, the quarantine period passed, the bald patch grew hair again, the parasitic mange, (if it ever existed,) was cured, and we finished the job. In the end the move to Grasmere turned out to be a good move for other reasons - we suddenly had a social life. Grasmere was buzzing with people our own age, writers, musicians, dancers, activists, spiritualists, climbers, movers and shakers. We met Derek Hook of the Harvest Vegetarian Restaurant in Ambleside (later the

---

2 *According to the DEFRA website, consulted in September 2012, the last confirmed case of Parasitic Mange was reported in South Glamorgan in 1977, so either they eventually decided that Ginger was falsely accused, or they just wanted to forget about the whole thing.*

proprietor of Zeffirelli's) and we hung out with the chilled-out residents at Allen Bank, where de Quincy and Coleridge had hung out with Wordsworth. We met Taffy Thomas, then a top dance caller, (now a famed storyteller) and Chrissie the dancer who ran a dance studio and later became Taffy's partner. On wet days we hung out in Fred Holdsworth's Ambleside bookshop, where the genial proprietor would be happy to talk poetry, order esoteric books about how to save the planet and keep us supplied with good coffee. We befriended Dave and Sylvia Hicks, and their boys Jay and Matt, and talked green politics and Peace Studies while drinking wine late into the night at Whitemoss Common. Aya ran a series of spiritual events at *Ishvara,* and I was called upon to drive Sir George Trevelyan on a sight-seeing trip around the Lakes. Suddenly we were back in the world after a long spell in the wilderness.

In particular we encountered Pete and Mags Laver who worked at Dove Cottage, Wordsworth's house, where Pete was the librarian. Pete was a poet, artist, wit and *bon viveur* and Mags was his witty and cynical foil, and they had the liveliest friends and wildest parties we had encountered since leaving my dad's farm in Lancashire. We had been living out in the woods for over two years, and had missed out on a whole era of poetry, TV, movies, music, newspapers, the early days of punk, pop culture, and all the rest. After my monkish abstinence from the media, suddenly I could not get enough. I fell in love with Debbie Harry within about 30 seconds of seeing her sing *Heart of Glass* on Pete's TV. (We had not seen a TV for about two years.) With Pete and Mags we caught up fast, they introduced us to punk, and their company kept us sane though the nightmare of the parasitic mange episode. One minute they would be discussing marginal notes in Wordsworth's manuscripts, and the next minute playing bawdy party games.

One day Pete went to the (infamous) Grasmere GP doctor, complaining of deafness. The doctor looked in his ears and said "Aach – there is enough wax in there to block up the arsehole of the Sphinx!" Another time Pete complained of chest pains and the doctor said "There is nothing wrong with you – go home and beat your wife!" Four years later, in 1983, aged 37, Pete died of a heart attack.

Penny Rock wood was also good for business. Jon Williams of the National Park now knew that we were serious, and he knew that we could deliver, so he promised us more work. Our time with The National Park also brought us some press attention - Penny Rock was, and is, a popular public amenity area, so we had a constant stream of visitors, walkers, photographers and journalists. *The Westmorland Gazette* ran a feature, and I was invited to submit a long article for *Horse & Driving*, a glossy and informative magazine read by both the heavy horse fraternity and the competitive driving world. The magazine title did not survive and it soon ceased publication, but at the time it was indispensable, and I spent a few happy evenings putting my article together, to be rewarded with a double page spread, with photographs. It was published in August 1979, and brought a little flurry of enquiries from enlightened woodland owners.

Penny Rock was another break-through, and a vindication of our optimism. We no longer needed to justify our existence by environmental idealism alone - we could compete on merit, on sound business planning and good silvicultural principles. We were still as poor as church mice, living hand to mouth, but we could have a good life while we were doing it.

14. West Cumbria
- landscape with horses

The oak butts from Penny Rock Wood were loaded on to a wagon as the snow melted, and our bureaucratic nightmare with the Ministry of Agriculture veterinary service became another statistic. We had a tearful farewell with the ever-generous Aya, who now had a gentleman companion, so we left the Aga and the rabbits for him to deal with. We walked Ginger and Charlie back over the rickety bridge and back over Red Bank to Colwith, then packed up the tin trunks, and moved back to our caravan at High Park.

We were soon on the road again. The National Trust were keen not to lose our services to the National Park, and to keep us working they offered us a job on selective thinning of conifers at Holme Wood, which rises steeply up the hillside on the west side of Loweswater in West Cumbria. We travelled out there to take a look with Ken Parker who introduced us to the head woodman at the National Trust sawmill, William (Billy) Armstrong and his wife Edie. Billy showed us a small plantation nearby where we could park the caravan, and as the spring came we shifted again. The National Trust warden Peter Taylor provided a landrover to move our rig, and we sailed over Dunmail Raise in fine style. We dragged the caravan as far we could get it up the track into Scale Hill plantation, so that we were just out of sight of the road, then went back to make another trip to collect Ginger. Charlie went back to his herd of mares on the hills above Whitworth in the Rossendale valley, and another page turned.

Billy and Edie Armstrong took a friendly interest in our welfare, and frequently invited us in for what they called 'tea'. West Cumbrian 'Tea' is a specifically Northern meal, sometimes called 'high tea' - different from dinner, a bit closer to supper and world away from the Southrons' 'tea' which comprises cucumber sandwiches and digestive biscuits. We would sit with them as the evening twilight settled on their ancient and immaculate sitting room, where they treated us to an enormous spread of home made bread, pies, rolls, cakes, scones, tarts, jams and jellies, and a great dish of poached salmon. (Billy was the water bailiff as well as head woodman for the Trust, and more often than not there would be a huge salmon hanging in his workshop next to the chainsaws and tractors.) "Reach up!" was the Cumberland equivalent of 'Help yourself!' and we did not need asking twice.

At Scale Hill – fixing the Commer Karrier

Ali took a job as a chambermaid at the Scale Hill Hotel, making up the beds for the guests, among them the comedienne Joyce Grenfell. The owners of Scale Hill allowed us the occasional use of a bath, and our farming neighbours Phil and Julia Walling would invite us for splendid farmhouse dinners and countless bottles of wine. Phil loved a lively debate and had strong opinions, and he soon gave up farming to become a barrister and an author. Julia baked cakes to die for, and looked after their three precocious and charming children, so those were memorable dinners. We would float back home in the small hours, full of wine and adrenalin, across the frosty fields to our cosy caravan where the moonlight was so bright that the shadows in the frozen wheel ruts looked to be made of black paint. We were having the time of our lives, and we knew it.

We became West Cumbrians for the summer. Ali's work as a chambermaid paid a proper wage, and each day after feeding the chickens and putting the bread on to rise, she would walk up the lane to the Scale Hill Hotel, watching the wild strawberries in the hedgerow come into leaf, then into flower and then into fruit. In June she made a tiny pot of wild strawberry jam. Each day I would take the Morris Traveller off to Holme Wood above Loweswater and pull trees out of the gloom into the sunshine, and at weekends we would drive out together to feed Ginger and enjoy the lonely and spectacular scenery.

Holme Wood and Loweswater, Grasmoor beyond

Because of the extraordinary peace to be found there, the wildlife around Loweswater thrived. There were fish-eating birds taking advantage of the fish in the lake, and in the woods were owls, buzzards, red deer and badgers. The lower slopes were mostly mixed broadleaf woodland, and shafts of sunlight would reveal glades of bluebells, wood anemones, wood sorrel and violets among the ferny, leafy dampness. There was the finest, most delicate of waterfalls, where a crystal beck tinkled over a wide rocky shelf covered in moss, like a Japanese painting, and it became my favourite place to eat my midday bread and rest after the exertions of the morning. I had been interested in early Celtic nature poetry since Arvon Foundation days, but it was in Loweswater that I first understood why the Bronze Age Celts and their shaman priests, the Druids, believed that access to the divine was best attempted in an oak wood, beside running water, with sunbeams glistening off a gentle waterfall.

Our base camp in Holme Wood was the old stone bothy by the lake shore. It had once been a fish hatchery, and later a bark barn, used to store the peeled oak bark used for traditional leather tanning. The bothy was ancient, dark, damp, and black with smoke, but more importantly it was a mile from the nearest neighbour, and the little rocky bay by the front door was home to herons, mergansers, great crested grebes, coots, tufted duck, mallard, goldeneye ducks and pochard. The view from the bothy doorway was breathtaking – Melbreak, the magnificent Grasmoor, and Whiteside, usually rising out of the mists, often illuminated by lightning and occasionally glowing with that extraordinary combination of purple heather dissolving in the sunshine into to the greys, browns, greens and yellows of scree, grass and bracken. To see the moon rise over Grasmoor could make me weak at the knees.

In early June the routine of horse work and Ali's hotel work was interrupted by our annual pilgrimage to Appleby Horse Fair. We boxed Ginger down to my dad's yard in Rossendale, and with Walter and a small posse of friends and family, we set off once again on the 90 miles road-trip to Appleby. This time we took five Fell ponies plus Ginger, with two bow-top wagons which Walter had built himself, and the same governess cart that we had used to explore North Yorkshire.

Our party was made up of my father and his new partner, Gill Barron, with her children Tom and Mike Barron, and my brother Tom. We travelled with the musical crew we had met four years earlier at the Arvon Foundation, when they were living in the cottages at Lumb Bank. After they were evicted they moved to Rossendale where they camped in Walter's pen on the common land at Shawforth. Pete and Dave were top session players of the time, with a repertoire of flying Irish jigs and reels that ensured a jumping pub session and a lively night round the fire, so with their partners Geraldine and Yvonne and children Sam and Bindle, plus my dad's new family, we had lively time. Ginger pulled one of the bow-top wagons, and Echo, the Fell pony stallion, pulled the other one, helped by a Fell sideliner called Rhino. We managed 30 miles a day with the usual routine of loose tyres, loose axles, loose horses, and more than likely a broken shaft.

Bill, Ali and Tom Lloyd on the road to Appleby Fair 1979.
The ropes (bottom left) attach a 'sideliner' to the wagon,
to steady him down and get him used to pulling.

Because Ginger was a powerful horse we were able to take a route which we usually avoided, via Forest Becks near Sawley. The hills on the route were too steep for the small Fell Ponies and a loaded wagon, but we knew that Ginger would manage them so we transferred the heavy items from Walter's wagon into ours. To slow us down on the steep descent to cross Forest Becks we used a heavy iron drag shoe as a brake. The drag shoe is chained onto the body of the wagon and then placed under a rear wheel to prevent the wheel from turning. The shoe skids along the ground, and the friction slows down the wagon and takes the weight off the horse's britching so he is not pushed downhill. The wagon was heavy and the descent was long, and by the time we got to the bottom, the drag shoe was so hot that it had melted the tarmac and made a groove in the road fifty yards long. I still pass that way from time to time and last time I looked the groove in the tarmac was still there, after 40 years!

Appleby Fair is unruly, unpredictable, and exciting. In 1979 there was no Market Field, and the traders set up their stalls on Fair Hill, which was tightly packed with caravans and horseboxes. That year being the first time we had travelled 'by road' (i.e. horsedrawn) with Ginger, we carried a heavier load of pots and pans, firewood, hay bales and hard feed, so we could make a proper camp and felt as if we belonged. I took my brother Tom swimming in the River Eden on Ginger's back, then we made a stick fire, opened up the beer, and took out the instruments. Soon there was a good crowd to listen to the music – Big Roy danced with a horse, holding its front legs in the air, and Tom Walsh flaked out the tunes on his melodeon, I played the whistle, Ali played the fiddle and my dad played the drum. A passing photographer took a great shot, and came back the next year to give us a copy.

As we pulled off Fair Hill, our friend Colin Butterworth presented us with a pair of Old English Game birds. We named them Dermot and Grainne after the Irish legendary heroes, and took them with us back to Scale Hill, where they roosted in the branches or on top of the caravan, and Grainne laid the occasional egg.

My dad sold his bow-top wagon to Mr. McMeakin from Dalton-in-Furness for £900 and I claimed some of that for delivering it on my new

Appleby Fair 1979. Tom Walsh on melodeon, Walter on drum,
Big Roy dancing the pasadoble with a pony.

(vintage) truck - an ancient Commer Karrier. Our friend Bobby Jeffryes boxed Ginger from Appleby Fair to Mockerkin in West Cumbria, and I walked him back over the hill to Loweswater, only to find next day that he had developed strangles, a common and highly infectious respiratory disease, and always a risk at Appleby Fair where he mixed with 1000 other horses. The vet recommended that he be kept off work for three weeks. Ginger recovered, but meanwhile I was without a work horse.

Luckily I had made the acquaintance of Joe Huddart, a general dealer and horseman who had a timber yard in Cockermouth. Joe would send his lads to work his horses in the woods to get fence posts and firewood for sale from his yard, so I knew he kept a couple of useful snigging horses. He agreed to lend me one, by the name of Sally, in return for me getting him a load of wood for himself. I tried her out, and she was good enough. Not as good as Ginger, but she was quiet and strong and willing, and she kept me in business until Ginger was fit again.

After Holme Wood was done we walked Ginger all the way up the valley to Crag Wood at Buttermere, not far from the famous Buttermere

Pines near Gatesgarth Farm, to extract larch first thinnings in an isolated wood in the most magnificent landscape. At Buttermere I worked on my own – Watto was too far away back in Ambleside, and Ali was working at Scale Hill Hotel, so for the whole of that summer I would make the trip up the valley each morning in the Morris Traveller, haul a few tons of wood, and cruise back down at night. I grazed Ginger nearby at Gatesgarth Farm, courtesy of Mr. Thomas Richardson. He was a tall, grizzled, friendly man, interested in horses, and he offered me a job using Ginger as a pack-horse to carry fence posts for a new boundary fence, high up on to the fell. He took me up the fell to show me the way as far as we could get in his Landrover, until we reached the narrow paths and rocky steps that no vehicle could manage in those far-off days before ATV bikes.

I needed a packsaddle, so I travelled to Rossendale again to see Walter, who knew all about packsaddles and even built them himself. He had a collection in the same barn we had converted into a theatre several years before, and there I found a tubular-steel military item that had once

Appleby Fair. Washing the horses in the river Eden (Photo: Lee Earnshaw)

been used by British army mules to carry artillery over the mountains of Malaya. With a bit of work I adapted it to fit Ginger, and found a way to load him up with fence posts instead of gun barrels. We could take about a dozen posts, or else two rolls of sheep netting at a time - the tracks were so steep and the going so rough that a dozen posts was a big load - but we managed four trips a day. Ginger was too big for the job - a smaller Fell or Dales pony would have managed better, but we did it. The climb was exhausting, and the rate of pay was pitiful, so we did not look for more of that kind of work, but we took some pride in it, and I hope the fence is there still. If it needs replacing again, then an ATV quad bike would do the whole job in a couple of hours, but the ATV rider would miss out on the exhilaration of a slow and peaceful climb onto the high fells, and the immense relief and satisfaction from laying down to rest on the top.

The landscape of the Lake District has inspired so many writers for so long that it is difficult to describe without falling into cliché. There was no doubt that landscape sustained us throughout our adventure, and gave our lives beauty and another level of meaning and purpose, and Buttermere was the peak experience. I would drive up the valley in the early mornings, around the road blocks of sleeping Herdwick sheep, through dramatic combinations of mist, sunlight and shadow. The deep valleys produced extraordinary optical effects, especially at dawn and dusk when the light moved from a calm soft blue tinge to glorious golds and yellows. In those brief 'blue periods' the purple and white of the heather and the farmhouses on the fellside would glow as if emitting their own light, while on the dark and cloudy days the fells would appear monstrous and threatening as they were glimpsed though the mist.

In the evening, however hard the day's work may have been, however little timber had been extracted, however many times the chainsaw broke down, or the horse lost a shoe, however pitiable the bank balance, and no matter how many times I reflected on how foolish I had been to pursue this ideal so single-mindedly while my contemporaries were making huge salaries and driving flash cars, no matter if I sometimes told myself that I had made a seriously wrong move, the landscape would lift me out of my mood. Particularly at the end of the

At Scale Hill

day, weary, sometimes bruised and frustrated with the struggle, I would come round a corner and see the lakes and fells spread out in front of me in a spectacular vista. The Lake District may not be dramatic in scale like the Alps, but it has a breath-taking and subtle beauty. The effect comes from a combination of ancient, ancient rocks, glaciated and worn into steep and rugged profiles with shining screes, covered in rich lush green vegetation, giving way up the slopes to bracken and heather, decorated by glittering, or darkly brooding lakes, shimmering tarns, surging rivers, tumbling becks and cosy little whitewashed houses. And of course it was dotted with the magnificent woods where I spent my days, each of which was a new world.

Even if we were caught up in a hopeless and misguided ideology about sustainability and horsepower, we felt blessed and privileged to be able to live and work in such a landscape. In the words of Walt Whitman, we embraced *"the bracing and buoyant equilibrium of concrete outdoor Nature, the only permanent reliance for sanity of book or human life."* The drive to and from the woods at Buttermere was unforgettable, like the view of Coniston Old Man coming home from Harrowslack, and the view of the great mass of Wetherlam on our first sparkling dawn at High Park.

The continued existence of this landscape owes an enormous amount to the National Park, the National Trust, the Forestry Commission and the agri-environemnt schemes run by DEFRA. These agencies make it possible for the hundreds of hill farmers to maintain the landscape of the Lake District fells and valleys using mostly traditional methods. Although I have had my differences of opinion with all of them at one time or another, the extraordinary beauty of the English Lakes could not have survived without the enlightened policies of these organisations, and it gave meaning and purpose to our lives to be making a contribution to that beauty.

In September 1979, Ali and I were married in Cockermouth, and our friends Peter and Kate came from High Park to witness the ceremony. We preferred not to have a party, to the *chagrin* of my family, particularly Uncle George who loved any opportunity for a family get-together. They all lived too far away to make a trip to West Cumbria, we had no money to spare for a wedding party, and anyway we preferred our own company and that of a few friends, so after signing the book we retired to the caravan for tea and fruit cake with the witnesses, skipped the honeymoon, and carried on as before. Eventually my mother persuaded me that we should have a small celebration at her house near Hailfax, and we made the trip, drank champagne, and dined at a proper table. My mother took a rare picture of us playing the Octopus Reel on a pair of tin whistles.

When we found out that we were expecting our first child, we knew that everything would change. We loved our life in the wood, but we decided that it would be too hard to raise children in a caravan - not enough time, not enough money, not enough space in the caravan, too isolated from our friends and family, and just too dangerous, so we decided to quit the horse logging business and find a proper house and a better rate of pay. We told the National Trust that we were quitting, and they were so keen to have us stay in the woods that they immediately offered us a cottage in the village of Near Sawrey, the home of Beatrix Potter not far from Hawkshead. How could we refuse?

We accepted without a second thought, but after several months of waiting we were only a few weeks away from the expected birth date and still the promised cottage had failed to materialise. I phoned the National Trust and left messages, but no response. In desperation I wrote a rude letter to their land agent, telling him that we had had enough and suggesting that he pull his finger out. I heard later that he was enraged at our cheek, and I suppose I was a bit snotty, but I had lost patience with broken promises, and I thought we deserved better. The rude letter might have been disaster, but it turned out well in the end – we moved in to Bankhead Cottage, Near Sawrey, on 15th November 1979.

Bill and Ali playing the Octopus Reel at their wedding party, 1979

15. Near Sawrey – new roots

Bankhead Cottage was everything we needed, and more. Known locally as 'Seldom Seen', it was a typical late 19th century semi-detached estate cottage. Bequeathed to the National Trust by Beatrix Potter, it was tucked away on a hillside about 500 yards from Hill Top in Near Sawrey, but almost hidden. Most of the land around belonged to the Trust, and Geoff Storey of Hill Top Farm was happy to rent us a field up the track to Moss Eccles, where I erected yet another ramshackle shelter and hay barn.

Bankhead Cottage, Near Sawrey. Our house, No 2, on the right.

In my collection of battered Victorian farming books we had a wonderful volume called *The Complete Grazier,* given to me when we were clearing out Brock Well, near Halifax, a derelict house belonging to a distant cousin. The book was an encyclopaedia of farming, and had a whole chapter on how to build a farm-worker's cottage. Bankhead Cottage was straight out of the manual. The front door opened onto a sitting room with a staircase, and behind that was a kitchen with a back door and enough room for a sink and cooker - and two tin trunks full of harness. Upstairs were two tiny bedrooms and what had been a third tiny bedroom, now a bathroom. Outside at the back was the old earth closet toilet and a wood store. Out front was a slated porch and a small garden (about the size of four blankets) where we built a compost heap and Ali planted potatoes and vegetables so crammed together that they were difficult to harvest, although they tasted just right. Up the back was a rough enclosure with a few ancient trees, a drying green and a chicken shed. Next door were Stan and Sue Barr, the best of neighbours. Stan introduced himself straight away, and just before Christmas he invited me round to his house for a festive whisky. The next morning (did I even go to bed?) I had the first, and what remains the very worst, horrendous whiskey hangover of my life. Ali had to drive us both south to Halifax for the family Christmas, heavily pregnant, tight-lipped all the way. (*Whisky you are the divil)*

Stan Barr was built like an Ardennes horse, and the strongest man I ever knew, stronger even than Mark Morton. Stan was a champion Cumberland wrestler, a clever boxer, a professional rifleman and licensed deer-stalker. His hobby was explosions - mostly gelignite and 'plastic.' He kept guns, dogs, pheasants and bees, and a wonderful electrical blasting trigger device called a *Beethoven Mark II Exploder*. He drove a JCB digger for a living, and could handle it like a knife and fork. The stories about Stan Barr would fill another book, particularly those stories told by Watto, whose farm at Sunny Brow was near to where Stan and his brother Billy Barr were brought up.

Stan shared a hobby with my father - accumulating unserviceable items, too good to throw away. Some time after we moved in to Bankhead and added our own collection of trailers, vehicles, piles of firewood, stacks of hay and old tin trunks, our landlords the National Trust wrote us both a stiff letter complaining about the 'awful scrow' that had accumulated outside Bankhead Cottage, and in particular the scrap cars. Most of them were towed away, but one without wheels was too far gone to move, so it was simply buried in a hole and covered over using Stan's JCB - but only after he had nonchalantly removed a box of gelignite and a handful of detonators from the back seat.

Bankhead Cottage: Stan Barr, burying a car.

Stan Barr was handy with explosives, and whenever any blasting was needed, to clear rock for a building site or to make a new waterway, Stan would be engaged to do the job. I asked him one day if he would use his digger to level a platform at the back of the cottage, so that we could put up a wooden shed for storage, and he decided to use gelignite instead of the digger. He laid a stick in position, added some 'plastic' for fun, inserted the detonator, and ran his wires back to the 'Beethoven Box.' We took shelter under the porch roof, wound the handle to charge the capacitor, and pressed the button. The blast was heard at Outgate, four miles away, and the noise of rock clattering down on the house roof sounded like a machine gun. "I must have used a bit much", said Stan, grinning like a monkey in comic book style.

Within a week of moving to Sawrey, I started work on selective thinning at Claife Heights. At first the National Trust gang did all the felling, but before long I was marking and felling my own trees, and negotiated a better hourly rate. Because I was now a National Trust tenant, and 'on the books' I was confident enough to write to them and ask for a raise. After all, we now had another mouth to feed and rent to pay! Ken Parker was sympathetic, and I backed up my request with a breakdown of my costs and the value of my work, based on two years' actual hands-on experience, with a set of accounts to prove it. Ken agreed, and at last I had a decent hourly rate.

As part of my deal with the National Trust I was given the use of the old plank barn at Harrowslack woodyard next to the sawmill on the west side of Windermere near Ferry House. The Harrowslack woodmen were old-school, friendly and helpful and they took to Ginger as if he was their own. They fired up the stationary diesel engine to drive the huge moving-bed circular saw in the sawmill and used it to cut a dozen larch butts into planks, and next day they boarded out half of the inside of the old plank barn, installed a feed rack and cleared out the other half as a hay and feed store. They provided bags full of sawdust for bedding, and proudly produced photographs of themselves as young men, working their own horses in the days before tractors. Thanks to the camaraderie of horsemen, for the first time I had a stable as well as a house to live in, and it was luxurious to yoke up in the morning by the light of a hurricane lamp when Ginger and I were both clean, dry and well rested.

In 1980 our daughter Eleanor was born and I worked the woods at Harrowslack and Claife Heights on and off for the whole year. The Claife woods were varied and interesting, with tremendous views over Lake Windermere to the north and east, back towards Coniston and Wetherlam to the west, and on a clear day you could see the foot of the lake and out across Morecambe Bay. There was a green track leading up to the Heights from the water's edge near Ferry House, and each day I would ride Ginger up the track, through a tall plantation of European larch, and onto Claife Heights.

At that time I was playing the banjo as much as I could (tiny infant permitting) and I had a routine for learning the words new songs. I would write out the lyrics, keeping the paper in my pocket when I went to work. Before starting the climb up the hill from Harrowslack onto Claife, I would memorise a couple of verses and then repeat them to myself as we climbed. After a while I would sing them out loud, and before long I would be singing with full voice. I still sing many of those songs 40 years later, and whenever I sing *'Wasn't Born to Follow'* or *'Ballad of Easy Rider,' 'Glass of Wine'* or *"Windy Mountain',* my mind goes back to the days of riding Ginger up through the tall larch woods, on the long slow climb.

Selective thinning on Claife Heights, above Lake Windermere

After two years on the road, we were happy to put down some roots. We had a telephone for the first time in 4 years, I was becoming better known as a singer, and our network of musical, social and environmental activist contacts was growing. I had visits in the wood from Rod Everett of the Middlewood Permaculture Centre, Edward

Acland, environmental activist and tool guru from Sprint Mill, David Hicks from the Peace Studies Centre in Bradford, Michael Gee, landscape planner from Old Brathay, Andrew Morton the horseman from East Yorkshire, Lewis Cleverdon, the wheelwright, and George Van Weinen from Pembrokeshire. Ric the shepherd came from Dumfries, George and Christine Tardios from the Arvon Foundation, and even some old friends from the University Drama Department. Mary Barratt the otter spotter was curating an exhibition about rural crafts at Holker Hall and wanted a section on wood snigging for her display. Carolyn Francis the fiddler was working at Lakeside and Cat Crag and we would get together for fiddle and banjo sessions. The Lancaster Clocks Back Traditional Music Festival was starting up in the Trades Club, and every Sunday I played folk music with Steve Grundy in the Cox'ns Cabin in Bowness. The Ireby ceilidh was too far away to travel, so we started our own ceilidh band, Devil's Gallop.

Bill and Ali outside the stable at Harrowslack, with our new signboard.

Between jobs on Claife Heights I worked for a few weeks thinning larch at High Cross at the north end of Coniston Water with a National Trust chainsaw gang where I had a visit from the renowned Bill Hogarth who happened to be working nearby. Bill was the last of the full-time coppice workers in the Furness area, and 20 years later his work inspired a whole new generation of young woodland pioneers when he gave his name to the Bill Hogarth Memorial Apprenticeship Trust. He needed a big horse for an extraction job at Force Forge, in the Rusland Valley, so I went to meet him and he showed me the job - second selective thinnings in a commercial birch plantation. The poles were tall and straight and were to be sold for turnery, so they had to be cut to length and then 'striped' - i.e. the bark cut through into the sapwood with the end of a chainsaw along the whole length of the pole on four sides, to let the air in and speed up the drying process.

I was keen to work for Bill Hogarth, even though it was back on piecework, partly because he was an old-school craftsman, and partly because he promised regular work, and valued the work of the horse in coppice woods. I reckoned that it did no harm to show the National Trust that I was in demand, and if they wanted to keep me they would have to look after me. With hindsight I need not have worried because in fact they looked after us all very well, and they knew at least as well as I did that forestry horsemen were a rare breed, and they owned thousands of acres of Cumbrian woodland that were ideal work for wood horses.

I fixed my piecework price with Bill Hogarth, and the day after I finished the job at High Cross, Ginger and I were on the move again. It was half a day's walk from High Cross to Force Forge, winding through Grizedale Forest, and Ali and infant Eleanor came to collect me to get me home. The birch wood at Force Forge was a perfect job for horses - open and airy woodland, a mixture of gentle downhill and level going, with little brash and no drain gullies. I was weary of gloomy spruce trees, and even the larch plantation at High Cross had been dark and shady while the needles were on the trees in the summer time. The birch poles at Force Forge were big enough to pull one at a time, and even the biggest ones were manageable, so there was less effort in gathering a load, and because the poles were long and slender they stacked well. There was not much conversion work - one chainsaw cut to remove the top, another cut for the firewood, and then the long stripes down the butt for the turnery wood, for which I was paid extra. Dressing out could be done with a billhook, which I also used to mark the trees for felling. I did my own cutting, so I could plan the work to suit myself, and at the end of the day the lugubrious Henry from Force Forge farm provided stabling, entertainment and hot tea.

Long larch poles at High Cross

As soon as I had 20 tons birch wood stacked in the loading bay, Gilbert Brown's wagon would come from Hawkshead and take it away for making rolling pins, brush handles, bobbins and wooden spindles. Everything below four inches diameter went for firewood, stacked in length in a separate heap. Unusually, Bill's main interest was in the tops of the birch trees, which he needed for making his besom brooms, or for sale to Cartmel Race Course, where they were used to make the steeplechase jumps. I would take the top off each tree in one cut, extract and convert the pole, and then drag the whole tree-top out of the wood to the loading bay where they took up so much space that Bill had to get them away every few days.

Force Forge was a good contract, but it was the scene of another accident and this time I was lucky to escape alive. At the northern end of the birch wood was a high voltage power line, and along the power lines grew some of the tallest trees in the wood, most of which I had to take out. Although the National Grid contractors came regularly to clear the undergrowth below the power lines, they did not cut back beyond about 10 feet on either side, and the nearest trees were taller than the lines. I was well aware of the danger in felling trees close to high voltage lines, but foolishly I thought that it would be safer to do the cutting when there was a North wind blowing, so that the trees would be certain to fall in the right direction for extraction, and more importantly, away from the lines.

I waited for a day with a good north wind and started to fell along the line. The first few trees came down according to plan - I was by then experienced enough to drop more or less anything my saw could handle, more or less exactly where I wanted it, - or so I thought. After the first few trees were down I became overconfident, and did not pay attention to the wind, which had become gusty and changeable. After cutting the gob on one big tree, I made the back cut, carefully checked the angles to get it where I wanted it, and then made the final felling cut. Just as I cut through, a strong gust came from the south, and dropped the tree backwards, directly onto the power lines.

As the tree fell, it twisted in the wind, and pinched the chainsaw in the back cut, so I could not clear the saw as it was falling, and I still had both hands on the saw trying to work it free when the tree hit the wires. It all happened very quickly, within two seconds, and luckily the tree was rolling as it fell, so the top of the tree slipped along and then rolled off the lines. The lines did not break, but I saw a row of huge yellow sparks jumping from the wires, bouncing down the trunk towards me. Next thing I knew my feet left the ground and I was moving backwards through the air, pushed with surprising force. I felt my eyes bulging out of my sockets as time slowed down. I landed on my back, about 10 feet away from the stump, luckily without the saw. I am still unsure if it was the tree or the electricity that pushed me, but it felt like the force of a high voltage. Because I had my rubber wellies I was not well earthed, so there was a low current, and I survived. I had no burns, and no injuries apart from a bruised backside, a twisted knee, and a powerful sense of my own stupidity, but it did make the hairs on my neck stand up and I felt decidedly shaky and unwell for an hour or two afterwards. I lived to tell the tale, but probably only just. Looking back it was mighty stupid, but at the time it was just a bad judgement, a wrong call. *(Please take care felling near the power lines boys and girls, especially in the wind!)*

Birch tops for Bill Hogarth's racecourse jumps

When Bill Hogarth's birch was done, I dropped on to my first job as a wood merchant on my own account, thinning a nearby beech wood for lugubrious Henry, and dragging it out to stack at the roadside opposite his farmhouse. I bought the standing wood from Henry, felled it, extracted it, converted it to one metre baulks, then had an anxious week or two fearing that someone would steal it until I sold it as firewood by stack measure to Sam Rigg at High Newton, who would not deal in cubic metres but only in the old measure knows as 'cords.'

While I was phoning around everyone I knew to try to sell the firewood, I called a contractor who was working very big oak in a steep and rocky wood at the northern end of Grizedale Forest, just a few miles away. He was extracting large butts with a double drum winch mounted on an ancient skyline crane, leaving the branch wood and tops to rot in the wood, as they were not worth the effort of extracting. Half the trees were on a plateau above a sheer rock face, with only a narrow path for access, and because of the terrain and the distance, he could not reach all the big timber with his tractor, hence the winch. He wanted me to use the horse to pull the butts in to a single winching point at the top of the rock face, and then after each load came down Ginger would drag the steel winch line up back up for the next load.

His plan was just feasible, because most of the dragging was on the level plateau which was inaccessible to his tractor and winch and not bad going for the horse once I got up there. The whole job looked a bit iffy because there was no grazing or stabling anywhere near, but it was on my route to walk Ginger home from Force Forge, and he was a new customer, I liked a challenge, and I needed the money, so I took the job.

The solution to the lack of grazing and stabling was to borrow a horse-box to use overnight. Each day after work I would back him into the box and provide him with enough water, hay and horse nuts until next morning. It was not ideal, as he did not really have enough room to lie down, but it was a short job and he took no harm for a few nights - horses often sleep while standing, and Ginger was an expert at that. I gave him a good two-hour break in the middle of the day so he could lie down if he wanted to, but he did not bother.

Even with the help of two snatch-blocks, pulling out the winch rope turned out to be the hardest work in the end - an uphill drag, pulling a heavy wire off a drum that seemed to have a sticky brake. Apart from that the job went well enough, and we finished in about a week, only to find that we had to wait about 6 months to get paid. I had been warned about this particular contractor, but because he seemed honest enough (they always do) and because the job was interesting, and nearby, paid by the hour and in big open oak wood, I had decided to take the risk. Bad payers were unusual, as most contractors needed their good reputation in order to recruit sub-contractors, but sharks exist in all trades, and horse-logging is no exception. I cannot now remember his name, or if he ever did pay me, but probably he did, otherwise I would still have his name engraved on my memory!

After that episode I was glad to get back to The National Trust, who were at least honest and reliable. Although they might sometimes make me wait a month before they paid me, the Trust was our mainstay - my main employer as well as our landlord. We had their cottage at a low rent, they had built Ginger a good stable which we occupied free of charge, and they found us enough work to keep us afloat so we could raise our family. It was as much self-interest as blind loyalty, but when the Trust offered me six months' work and a proper Victorian stable at High Close, on the top of Red Bank above Grasmere, I was mighty glad to accept - not least because winter was coming and I urgently needed funds to get myself yet another vehicle. I always needed funds to get myself another vehicle.

Bankhead Cottage and Esthwaite Water by Stephen Darbishire

Ali at Zeffirelli's 1981

16. Motors - DIY living

Readers who have no interest in things mechanical may wish to skip the next chapter. What follows is a litany of repair jobs on a series of vehicles, most of which we acquired just before they were scrapped. All except one were eventually duly delivered to their final resting place in the scrap yard, but every one of them was given a year or two of extra life. The litany is recited to explain why we did it that way, and to demonstrate two things.

(a) DIY mechanics was an essential component of our life in the woods, simply because using horses to compete with tractors could never generate enough income to have our motors maintained professionally. Our life seemed to be a constant struggle against machines, paradoxically trying to compete with them and trying to keep them running.

(b) Although we disliked combustion engines and the damage they did, we knew that it is more rational and environmentally friendly to run old cars (and trucks) until they are really and truly dead, rather than scrap them just because they break down and look a bit tatty. We just kept on fixing them. Menders Rule.

Stan Barr, our neighbour at Bankhead Cottage in Sawrey, had a good collection of scrap cars, which he kept for spares. When Ali and I arrived at Bankhead Cottage he drove a Range Rover (battered) for work and an Opel Manta (very fast) for pleasure, and would occasionally take me on white-knuckle rides to help him with the recovery of breakdowns, or the getting of firewood, or the clearing of drains in Windermere and Ambleside. It was Stan Barr who towed me home in my Commer Karrier truck after it broke a crankshaft on the M62 motorway. The truck was second only to my horse as my pride and joy at the time. It may seem odd that a book about the ideology of horse-power should have a section about motors, but we were never purists, and we regarded motors as an inevitable and necessary evil. Maybe in an ideal world, or possibly in the dark post-industrial apocalypse that we feared was just round the corner, there would be no motor-cars. But maybe not.

One reason we embraced the horse-drawn ethos was because we thought we were ahead of the game. We believed that the oil was running out, so neither the global economy nor the global ecology could afford the internal combustion engine, and soon they would all be obsolete. By using horses, we thought our future was probably more secure - a case of enlightened self-interest. How wrong we were.

In 1978, we certainly would not have believed any prediction of what has in fact come to pass by 2020 - the extraordinary growth of industrial processes, in particular car manufacturing, or the worsening planetary pollution, or the quantities of toxic waste poured into the oceans, the profligate waste of finite resources, the expansion of air travel, the laying of billions of tons of concrete and tarmac, and the relentless mechanization of vast areas of human activity. We would not have believed it, but it has all come to pass nevertheless.

We were not Luddites, and like almost everyone else in Britain, and whatever our ideology, we relied on the internal combustion engine to sustain our way of life. This might be considered hypocrisy, but we believed that we could compromise by keeping old motors on the road and doing our own repairs. That choice was the single most significant part of the precarious financial arrangements which allowed us to stay afloat while preparing for the post-industrial society, and went some way to easing our conscience about running motors at all.

For the whole seven years we were in the wood, I don't recall ever putting a car into a garage, except for new tyres and exhausts (which at that time came with 'free fitting') and for the annual MOT tests. I did change an exhaust system once, but only once. It cost me more to buy the part than it did to have it fitted free ... so we decided we could afford a professional for that job. Otherwise, we learned to do everything else ourselves, and saved thousands of pounds, which in any case we did not have.

We were helped by a great little book called *Truck* by John Jerome, in which he described in detail the practical, emotional and philosophical dimensions of stripping down a pickup truck, almost to its last component, and then rebuilding it. It was amusing, informative and inspiring. It simultaneously de-mystified and yet glamorized the mechanic's craft. It certainly prepared me for the inevitable banality which marked the end of an epic, difficult and complex all-night workshop session, from which I would emerge covered in black grease, and desperate for a cup of tea.

There is a particular mental, emotional and physical exhaustion associated with the cup of tea and the smell of Swarfega hand-cleaner which comes at the end of a long stint, in which an engine has been removed, dismantled, re-assembled and replaced. John Jerome knew this special exhaustion – his project took him a whole year, and he was very thorough. He described the blown air heater in his truck as 'like a mouse breathing on my foot,' and decided to fix it, so he set about removing, dismantling and rebuilding the unit, only to find that the rebuilt unit was not much different. That is how it is. He started out the year with an old, worn, functional, but slightly unreliable truck, and after a year of work he ended up with an old, worn, functional, but slightly unreliable truck, and a small heap of nuts and bolts that did not seem to fit anywhere anymore. (And a good book of course.)

After we left Littlebeck, the Morris 1000 Traveller did everything we needed, except pulling the caravan and the horsebox. Even if it had a tow bar, it would not have managed to pull a horsebox trailer with a

16 hand workhorse in it, so when the National Trust wanted us to work all over Cumbria, we needed a truck. Before we left Little Langdale I had acquired an ancient Commer Karrier Bantam, 30 cwt tipper, bought from Usher's builders in Coniston. After a hard life carrying building materials, it was already fit for the scrap heap when we bought it. It had been standing unused in a shed for a year or two after Ushers replaced it with a modern pickup, and was 25 years old in 1979, and a museum piece. The cab had rusted and rotted, the wooden sides were flapping about like wings, the engine blew a bit of smoke and it needed three layers of carpet in the cab to keep the noise down, but it had a Perkins 4 cylinder engine and it started first time - both highly desirable attributes.

I took a chance and bought it, somewhere above scrap price. The Ushers drove a hard bargain, and I probably paid £100 too much for it, but it was a darling truck, already a classic design, and I had to have it. We found a skilled man at Bowston who could fabricate a new piece to replace the rotted-out cab mounting, and when he had hammered it into place with a sledgehammer and welded it all up, we sailed through the MOT. Before long we moved to Loweswater, and I spent all my spare time on wet days and evenings cleaning, repairing, and rubbing down, and on a couple of fine days I re-painted it, (Racing Green and Forest Brown, of course) and then polished it until it shone like a showman's lorry.

The Commer Karrier could carry two people and 30 cwt (about what a modern 4x4 pickup can carry) but it had a tipper body and a top speed of about 46 mph. I loved it, and before long I was doing small deals and general haulage around West Cumbria - fence posts, firewood, rustic poles, and Westmorland roofing slates. I drove it to Morton's farm in East Yorkshire and brought back 20 sacks of organic wheat for delivery to the Watermill at Little Salkeld. I loaded up fence posts from the National Trust sawmill for delivery to farmers in Rossendale, then drove on to Morton's for a ton of quality seed hay for Ginger. I hauled a ton of firewood to Pendragon Castle at Kirkby Stephen, and never got paid for it. I was hired by Tom Timmins, General Dealer of Broughton, near Cockermouth, to deliver his various goods all over the North of England. He dealt in anything that might turn a profit, and had a yard full of bygones, horses, harness, carts and wagons, building supplies, hay and straw, and furniture, and he was sharp. One irate customer called round to see him while I was there, but could make nothing of Tom Timmins, and so he stomped off, scowling and unhappy. "I'll never deal with you again" he shouted as he left. "Nae worries" said Tom, "if I get round you all once, I'll be happy enough."

I would go trucking when the weather was too wet for the wood, and loved it, but it was too good to last. The Karrier Bantam was designed before motorways existed, and in November 1979, I pushed my luck too far. As we were about to move in to our new National Trust cottage, I used the Karrier for the last time, to shift our bits of furniture and possessions from storage near Halifax, where they had been for three years. The crankshaft broke on the motorway just north of Wigan.

That was a seriously stressful day, and I can feel my heart jump even now when I think of it. I was cruising at top speed (46 mph) when I heard the engine note change. I knew instantly that it was just plain BAD, but I had no money in the bank and no money in my pocket, so a motorway recovery was out of the question. Our household treasures were on board, and even though we had lived in the caravan without our house tat for three years, I was not keen to leave our precious load on a broken down truck on the roadside while I hitched home to arrange a recovery. So I just drove on, pretending it was not happening, not quite believing that the engine was still running and expecting terminal destruction at any moment. I got as far as The Tickled Trout, a landmark pub near Preston, where I could get off the motorway and get a tow home on the back roads. Even at 25 mph the knocking noise was getting steadily louder, but the engine was still running. By then I didn't care any more, so again I drove on. It seemed at the time like a miracle of faith, although in fact it was just good Perkins engineering, but I sailed on for another 20 miles, on past Lancaster, and by the time I got to Carnforth the knocking from the engine was so loud it was turning the heads of passing drivers. I pulled off the motorway and parked up in the Tewitfield lay-by, having covered 35 miles with a broken crankshaft, which taught me yet another invaluable lesson - never give up until it is over. I walked to a phone box, and Stan Barr came out in his beat-up Range Rover to tow me home.

The Karrier stood forlornly outside our new house at Sawrey until the warning letter from the National Trust, when it too went to the scrapper. That was another hard day, but I had done my best to save it. I removed the engine, had a second-hand crankshaft reground at Fell's Engineering at Troutbeck Bridge, and spent a fortune on new bearing shells which sat accusingly on our sideboard for months, arranged artistically around the re-ground crankshaft, gleaming like a modernist sculpture, but it ended badly. We had no workshop, no funds and no time to put it all back together. Our first baby had arrived, and winter was upon us. I was so busy trying to earn a living that I could not justify taking days off to re-build the engine. Before long I had added a new string to my bow and was supplying horses to film and TV companies, so I needed to travel the country with a box van and trailer, at more than 46 miles an hour. I admitted defeat, scrapped the Bantam, and bought an almost worn-out twin-wheeled Ford Transit for £400.00. It had 175,000 miles on the clock, but it sounded sweet, and it had the vital tow bar.

The Ford Transit and horse-box trailer at Bankhead Cottage

With the Transit I could diversify further. As well as animals for film and TV, I did garden jobs, tree surgery jobs and delivered rustic poles to garden centres all over the North of England. Because I was working on film sets and competing with heavy money, I had to smarten up, so I gave the Transit and horsebox a DIY paint job – Battleship Grey and whitewall tyres. I drove that rig all over the country, including a long slow trip to the horse fair at Priddy in Devon, but eventually the Transit died after failing to start one winter morning at Brathay Hall. I needed a tow start, and I crunched into the back of the towing vehicle in front when he stopped suddenly on the icy road. He had hardly a scratch, so he graciously made no claim, but his tow bar took out my radiator and front lights. The Transit was a write off, and the scrap-man had the last laugh.

Meanwhile, in the spring of 1980, we still used the Morris Traveller for getting to work and for our weekly shopping trip to Kendal, but although we had managed to keep it on the road against the odds for three years, it was not suitable for carrying heavy loads up and down bumpy forest tracks. Tom Pratt our neighbour at High Park had always said that a Morris Traveller was 'bad to kill', and we proved him right. All the maintenance and repair jobs on a Morris Traveller could be done by a keen (or penniless) amateur, so by the time it went though the pearly gates at the scrap yard, I had done just about every repair in the manual. With the Karrier and the Transit both gone, the Traveller was once more our lifeline. Bankhead Cottage was a long way from anywhere I wanted to go except the village pub, and that car had to be kept running at all costs, so when Rad-Weld could no longer seal the disintegrating radiator it was time to put on the overalls, get out the socket set for the *n*th time and fit a new one.

I had never had any training in auto-mechanics, but when faced with a broken down vehicle we had little choice but to fix it ourselves, with the help of the indispensable Haynes Workshop Manuals. I was initiated into the mysteries of car mechanics by the very first car we bought at Lumb Bank in 1975. It was a Vauxhall Victor, purchased from a scrap yard of course. It broke down very quickly, when the universal joint in the prop-shaft disintegrated in the middle of Hebden Bridge. We had no money to get it fixed in a garage, but our neighbour and drinking companion from

Lower Lumb, Michael Phillips, showed me how to fix it myself. With the Haynes Workshop Manual at the ready and a roll-up cigarette in hand, he had the prop-shaft off in 10 minutes. We made a quick run to the scrap yard to get a spare, and were back on the road within the hour. It was like a magic trick, turning disaster into enlightenment, and it made me realise that most mechanical jobs require only a little common sense, some decent tools and a workshop manual. Motivation is a factor, of course, and that comes remarkably quickly from a bank balance which is at, or closely approaching, zero. Encouraged by this discovery, I developed a weakness for 'collectible' cars. After the Victor was scrapped (failed MOT with flapping wings) I bought a Vauxhall Viscount.

At that time the Viscount was one of the largest cars on the road, inspired by American chrome-plated gas guzzlers. It had a 'straight six' 4-litre engine, automatic transmission (with overdrive,) birds-eye maple dashboard, real leather seats, electric windows, front bench seat, and a bonnet the size of a barn door. It could carry seven people and their entire luggage in gliding, whispering, silky luxury. The one I found also had a blown head gasket and a dodgy gearbox, but in my new-found enthusiasm and ignorance, I considered those a challenge. I bought yet another Haynes Workshop Manual, squeezed the car into the hay barn at Lumb Bank, lifted the cylinder head, replaced the head gasket, clapped the head back on again, reset the tappets and the timing, and wired it all up again. I then spent several hours trying to work out why it would not start. After a full day of fiddling and faffing, we eventually called the AA man. He removed the HT leads, replaced them in the correct order, and turned the key. It started instantly, with an unforgettable purring sound, and for month after month it started first time and ran like a Bentley.

The story almost ended when the automatic gearbox failed after a memorable fast road trip up into Wharfedale for a music session. I drew the line at fixing automatic gearboxes, but for the sake of the mighty Viscount, we jacked up the car, dropped the gearbox and took it to the leading automatic gearbox man in the whole of the North of England. He just happened to live in Blackshaw Head, about 5 miles away on the top of the high road to Burnley. His workshop-cum-garage was so far up in the hills that when we got there we found that the whole garage petrol pump was encased in a single solid block of ice. I could see the nozzle through the ice, but could not touch it. He fixed the gearbox, at eye-watering cost, but the Viscount failed its MOT a few months later with a rusted chassis which was so far gone that there was no solid metal left to weld to. It was a good car, as cars go, and as good cars go, it went – to the scrap yard.

After cutting my teeth as a mechanic on the Victor and Viscount, a Morris Traveller with a mere 1000cc 4-cylinder engine and a hole in the radiator held no fears! Over a few short years we had taken the cylinder head off more than once, replaced the gaskets, reground the valves, set the valve rockers with a feeler gauge, dropped the sump and replaced the big-end bearings, dropped the gearbox to replace the clutch, renewed the woodwork, replaced the back doors, re-wired the lights, stripped the steering box, re-piped and re-lined the brakes, varnished the woodwork and painted the body with green Hammerite. We fitted a new rear spring, (tricky that one), yet another new prop shaft, replaced the bumper and put in two new half-shafts. (The track to Bankhead Cottage was both steep and icy, and the Morris half-shafts so weak that they would break for fun.) We had poured Rad-Weld into the cooling system until it was like porridge, until eventually the radiator had more holes than metal.

I say 'we' did all this, and I mean 'we' – Ali became so used to the quirks of the Morris Traveller that she could diagnose them as fast as I could. At High Park Ali had the job of holding one side of the car up in the air with a long lever while I chocked up the axles on stands, so that we could drop the gearbox. At Littlebeck she helped to grind in the valves. When the electric fuel pump had a persistent but intermittent fault, she had to kick the bulkhead in front of the passenger seat to get it going again. On a long drive, Ali's foot would get sore from constant kicking, but she knew from the stuttering engine sound when it had to be done. On one of our regular trips to East Yorkshire to see the Mortons, the fuel pump failed completely, and the engine stopped altogether on the top of the hill between Kendal and Sedbergh. No amount of bulkhead kicking would get it running, so we had no choice but to push the car off the road, dismount the fuel pump, strip it down and try to find out what was wrong. It was a mass of fine copper coils and tiny relays, which we cleaned and reassembled by instinct and guesswork, and we managed to get as far as Sedbergh (downhill almost all the way) by kicking the bulkhead every 30 seconds, where I pulled into a garage to ask for some advice. We had not a penny to spare, and I told the mechanic I could not afford a new pump, or even afford to have him look at it, but I asked him if he had any suggestions as to what could be the trouble. He was not having any of it, and about as unhelpful as he could be. He thought, quite correctly, that I was trying to pick his brains, but if I couldn't pay him, why should he help me? I did not argue, but cursed him as we drove off, kicking the bulkhead as we went. We stopped again near Hawes, and had another try at repairing the pump. This time, by a mysterious process involving an elastic band, a matchstick, and some unfathomable good luck, it worked, and took us to East Yorkshire and back. When we got home I found a spare from a scrap yard, and it lasted the life of the car.

Looking back I still cannot see why the Sedbergh mechanic thought I was taking advantage of him, and I still think he might have been more helpful. We told him we could not pay him anything, so he would have lost nothing by telling us for free. Did he think I might one day steal one of his customers by offering to fix their fuel pump with an elastic band and a matchstick, instead of allowing them to pay him to fit a new pump? Or maybe he thought that I really had plenty of money in my pocket and was just trying it on? Maybe he needed his reputation as a hard man, or maybe he did not like hippies who wanted something for nothing, or maybe he was just bitter. Who knows? Who cares?

I make no claim to be holier-than-thou, but for me it has always been a matter of mutual benefit, not moral superiority. I had become used to the ethos of the horse-drawn travelling people, who will usually, almost invariably, help any fellow traveller on the road who needs help. Many times on the road with horses we needed a hand, and it was always there for us. In return we would always go out of our way to help any traveller who needed it. I have lost count of the times I have picked up weary hitchhikers, fixed other people's broken down motors, repaired their horse-drawn carts and wagons, towed them out of ditches, fed hungry hippies, directed lost walkers, and taken in waifs and strays. Once I even picked up a man covered in blood who had just committed murder, although I did not know it at the time – I thought he had been involved in an accident. These random acts of kindness cost me nothing, and often brought me unlooked for rewards that money could not buy. My willingness to help out someone in trouble is really self-interest – it is part of a reciprocal system – do as you would be done by – and for me it has almost always worked. It seems as if I have received as much help as I have given, although I never kept count. Sometimes I was abused, and occasionally robbed, so maybe I am a bit more careful now, but even after 35 years I still feel mystified by the refusal of a tight-fisted garage owner to help out a penniless young motorist by giving a bit of information. What goes around, comes around.

At High Close, with Ginger, Morris Traveller, and brother-in-law Peter.

So, back at Bankhead Cottage, with the Transit scrapped, we relied once again on the Morris Traveller with a hole in the radiator. I still had work to finish at Loweswater, and I had to keep the Traveller on the road to get there, so I splashed out and put in a brand new radiator. By then, just about every page on the Haynes Workshop Manual had oily finger-marks on it, and the Traveller was just too unreliable for the job we now demanded of it. In the end, I wrote it off by pushing it too far. While going down a forest ride at Holme Wood (a track never meant for any form of motor car) loaded up with gear and travelling too fast, I failed to see the spike of a tree root sticking up about six inches above the ground. The root spike entered the left hand longitudinal chassis member just under the engine, and exited just under the rear bumper. The whole chassis box-section had been opened up like a zip. I managed to limp back home by bolting an oak plank to the chassis, but that was the end of the road for the Traveller. I managed to sell it for spares to a West Cumbrian, who planned to swap the engine (and the new radiator) into another Morris 1000. It might be still running yet for all I know.

We needed another car in a hurry, so we bought a Morris Oxford saloon from a newspaper advert. It was larger than the Traveller, and I reckoned it was basically reliable as it had a similar engine which I could maintain if necessary. It was ultra cheap, and totally unsuitable. I ripped out the back seats to carry harness and feed, filled the boot with chainsaws and fuel, and it lasted until MOT time, when it predictably failed (barely legal tyres, blowing exhaust, loose steering, dodgy handbrake, holes in bodywork). Next please!

We heard on the grapevine that one of the National Trust officers was selling a worn-out Saab 99, for less than it would cost me to fix the Morris Oxford, so that became the first of our worn-out Saabs. They were a classic design, and great cars in winter, designed for Swedish snow and ice so they had great traction and mighty heaters, but a mighty good heater does not count for an MOT, and even classic designs rust out in the end. They were unusual, so spares were hard to find, and the scrap yard took it off our hands.

The second Saab 99 did not last long after I heard that a neighbour

At Harrowslack with Ali and Eleanor, and the Morris Oxford

was selling an old BMW 2002, in good condition, for £400. I could not resist. The BMW 2002 was another classic design, and drove like a dream – very fast, very comfortable, and with an 8-track sound system! (Thanks to that car I grew to love Dolly Parton at 103 decibels.) Out of respect for German engineering and my own lack of skill, I resisted messing with the BMW engine, which had a relaxed purr and smelled of hot oil like a racing car. We did replace a torsion bar with no tools except a crowbar, an adjustable spanner, and a set of pulleys – a job not to be recommended. After 2 years, the BMW went for spares to our local fish-farmer, Nigel Woodhouse, in exchange for a wad of cash and a box of fish, and I moved on to old Volvo 240 estate cars. The Volvo Estates proved themselves the very best cars for an impoverished forest horseman. They were built like tanks, with an interior like the Tardis, seriously reliable, and with lots of space under the bonnet for getting at the engine for the inevitable repairs.

Before the second Volvo 240 went to the scrap yard I found a Volvo P1800 sports coupe that had been standing in a garage in Ambleside for 20 years. I bought it as a non-runner, stripped the engine, restored the bodywork, rubbed it down, and re-sprayed it - British Racing Green, of course. By the time the last Volvo estate car had run into the ground, I had changed my job so did not need a workhorse vehicle, and ran the P1800 for several years, commuting into Kendal. My daughters loved the overdrive switch, which they were allowed to operate at the appropriate moment, while riding in the child-sized back seats.

I lavished care and attention on the P1800 but lost it to a negligent garage mechanic, who failed to install the battery clamp properly so the battery terminal made contact with the bonnet when it went over a bump. The next time I switched on the lights I had a spectacular display of sparks and black smoke, and a burned out wiring harness. The garage denied liability, and I could not prove otherwise. I had no money to fight a court case, so I cursed them to make myself feel better, and sold what was left of the P1800 coupe to the Keswick Motor Museum. The museum paid me peanuts, and I shed another tear for all my wasted work. So it goes.

Keeping old vehicles on the road long past their normal life expectancy cost us lots of time, some inconvenient breakdowns, some sleepless nights, and some skin off my knuckles, but they were part of the plan. My love-hate relationship with motor vehicles was a conscious choice, and not just lack of funds - there was also some idealism, pride and vanity, as well as practicality and financial necessity. New cars were expensive, but commonplace. Old cars and trucks, particularly classics from the scrap yard, were cheap, but also rare and special, and they turned heads. More important, by taking pride in good design and good engineering and rejecting the association of social status with new cars and built-in obsolescence, we could demonstrate that there was another way. Looking after the past became a habit.

Back in the world which we had abandoned when we took to the road in 1977, we might have expected to earn enough money eventually to run a reliable car, or at least to pay a professional mechanic. But in 1982, at Near Sawrey, we were not in that world. We were in our own world - a world that we had searched for and found, a world we had constructed for ourselves, a world of horses and woodland, a world of cottage economy, idealistic self-sufficiency and self reliance. We were trying to bring the traditions of the past to bear on the present and the future, and to establish for ourselves a new set of values for the post-industrial, post-materialist society which we believed was just around the corner. If those values meant I had to spend hour after hour stripping down engines, sharpening chainsaws, mending harness or cutting firewood, that was part of the plan. The hours spent digging the garden to grow potatoes, installing a wood-burner, building a shed or sweeping the chimney were not chores, but stepping stones in a stream, a bridge to a better future.

Ali and I were in agreement about how we should live. Once our children were born, Ali chose to look after the children, keep the chickens, grow the vegetables, preserve the fruit, bake the bread and make the clothes. She was, and is, very skilled. I repaired the machines, worked the horses and got the firewood. We both worked long hours to make it happen, and all our money was shared. That model is now called 'gender stereotyping' but for us it was not a 'social construct' imposed by the 'patriarchy'. It was cooperative, mutual, and freely chosen. Our gender roles were not exclusive, or tyrannical, and they suited our temperaments. We were willing to pay a price for our eccentric choices, because we believed that major social change was not only inevitable, but probably imminent. Only by DIY living could we sustain ourselves and our children when the lights went out, which we fully expected them to do.

Eleanor with the Volvo P1800 Coupe

17. High Close – *how it once was*

In 1982 we were five years in to our great adventure, and going strong. Our next horse project was a move to High Close, a Victorian mansion on Loughrigg at the top of Red Bank. The grounds overlooked Grasmere to the north, and Windermere and Wetherlam to the south. On a clear day you could see Snowdon 100 miles away in North Wales. The setup at High Close might have been constructed out of my best dreams. The Big House, surrounded by a few hundred of acres of woodlands and rough fell on Loughrigg and Langdale, had been left to the National Trust, who converted part of it to a Youth Hostel. At the front was an arboretum, full of rare specimens of trees brought back from exotic outposts of the Empire, and behind stood the old farmhouse and the Victorian stables.

At High Close, the present embodied the past, every day. Today was just like yesterday, and yesterday was 100 years ago. The stable yard was typical Westmorland architecture, with a hay barn at ground level, above an under housing at lower ground level, so that hay could be dropped into the stables and cattle stalls in winter. There were timber-framed balconies and slate roofed porches, with dressed stone drainage channels to carry away the surface water. The yard was both functional and pretty, but it was also an eccentric wonderland. Among the usual farmyard accumulation of tractors and implements, water tanks, heaps of wood and obsolete machines stood an enormous pale blue wagon with *Gavioli's Fairground Organ* and an address in Paris painted on the side in bold and colourful showman's designs. In the open cart-shed, the beams were just high enough to clear the funnel of a monster steam traction engine, with a painted name board, *Loughrigg,* all in full working order.

Ginger and I had the use of the old stables behind the main house. The High Close stables were mighty fancy, made for Victorian carriage horses, with cobbled floors, pitch pine stalls, and cast iron fittings - hay racks, hitching rails and feed mangers. The harness room had customised racks and hooks for each item of harness. For an aspiring horseman it was like stepping into a history book. The stables had not been used for carriage horses for over 50 years, and as I cleared out the junk I turned up ancient dusty gadgets out of time and place - an old wooden mortise lock, complete with wrought iron key, a set of huge wooden pulley blocks for moving enormous weights, a great glass carboy, a pair of wrought iron candlesticks, half a dozen giant horseshoes, a bag full of clog irons, and a box of worn out hedging mitts.

In one of the feed racks I found a cube of hardwood about the size of a bag of sugar, worn smooth and shiny under the dust, with a neat round hole through the middle. I recognised it from my Victorian horse manual as a halter-weight, which would be tied to the end of the horse's halter after the lead rein had been passed through an iron ring set in to the wall. As the horse moved his head up and down to feed, the halter-weight took up the slack in the rope so that the horse would not get tangled up in the loop as it moved around. I still have it as doorstop.

At the back of the stable was a connecting door to a feed room with great wooden corn bins for the horse feed and beyond that was the harness room. There was even a horseman's day room with a pot bellied stove, table and chairs, and a sofa for the inevitable wet days. I ratched the ancient jackdaw's nests out of the chimney, cleaned 40 years' worth of cobwebs off the windows, and brushed down the sofa. It took half a day of clearing and sweeping to get the place looking like a working stable, and then the snow came. I lit the stove and ate my sandwiches sitting in a time warp, warm and cosy as I watched the world outside turning white.

The neighbours were friendly. Zeke Myers who owned the *Gavioli* steam organ and the *Loughrigg* traction engine, lived in the farmhouse. Joe Park, an elderly retired National Trust woodman, lived in the cottage next door, and he would invite Ali and me in for tea and keep us entertained with his tales. The working horse was a great passport to conversation, and would often unlock the treasure-chest of memories of the old men who spent their working lives with horses. Joe Park had a good line in conundrums and tall tales, and although he was retired from the wood, he spent hours each day with an axe, bow-saw and wheelbarrow, gathering firewood from wherever he could get it and wheeling barrow-load after barrow-load back to his woodshed. I asked him why he still worked so hard, why he did not retire and take it easy. He grinned and said "coz I am a lazy man." I raised an eyebrow. "How so?" "Yep" he said "What I like best is to light my fire of an evening with one match. I want it dry enough to burn right away, and I don't like to have to keep getting up to blow on the fire." He spent his days collecting and splitting firewood, and stacked it to dry a year or two before he needed it. He would fill his log baskets each day, then all he had to do was to reach out from his armchair with one hand and put a log on the fire. He didn't even have to get up. "Aye, pure laziness."

The National Trust wanted us at High Close because they had three nearby woodlands to be thinned, all within half a mile of the yard. Although I worked on my own and the job progressed slowly, I saved them the trouble and expense of putting in a gang of woodmen, for which they had to provide a chainsaw per man, plus gang transport, a foreman, a tractor and winch, a workshop for tools and maintenance, and administrative staff to check the time sheets and pay the wages. The Trust had a backlog of woods waiting to be thinned, with not enough woodmen, so the gang was always busy. I was self employed, provided my own equipment, and needed no supervision as they trusted me to do my own marking, felling, extraction and stacking, and all on an hourly rate. Providing the five star stable cost them nothing, so the Trust had no overhead cost.

The first job at High Close was felling and extracting 20 tons of spruce from a difficult rocky site around a TV mast and that was all done in a couple of weeks. From there we moved to second thinnings in the larch plantation at Red Bank Wood, just across the road and overlooking Grasmere and Dunmail Raise. The wood was high up, a few hundred yards across a steep field, and the extraction route out of the wood was on a gentle downward slope through a gap in the wall into the field. I was felling my own wood so there were no high stocks and little brash, and the poles were felled in the direction of extraction. The weather was icy, and the snigging track became hard and greasy after a few days work, so the load would slide easily and Ginger could pull three or four poles at a time. Best of all there was a perfect site for a stack, with the right slope and all the right angles coming out of the wood.

Stacking is one of the frequent problems of horse-logging, because although the horse can drag poles, it cannot lift them. Most woodlands have limited space at the roadside, and the timber wagons will only come when there is a full load of about 30 tons ready to go. The hydraulic crane on the wagon (known as a HIAB after the manufacturer *Hydrauliska Industri AB)* has a limited reach, so the contractor must find a way of storing 30 tons of timber in a small space on the roadside where the wagon and HIAB can get it. It helps if the timber is converted to short lengths by cross-cutting, as the wagon can then drawn alongside the stack. Each pole can be cross cut and stacked immediately so that there is space by the stack for the next pole, but converting and cross cutting normally requires an extra man, otherwise the horse is standing around in the freezing cold between loads, which is not good when he is sweated up from working. But an extra man is very expensive if he is standing round waiting for the next poles to come in. Without an extra man, converting and stacking each pole as it comes out of the wood is hard on the horseman; after heaving baulks of timber up on to the stack he usually needs a rest before setting off back up the hill for another pole.

I was working alone, and I preferred to leave the job of converting the timber to the National Trust woodmen, but that meant I had to find a way of heaping up the poles so that the available space was not filled up in the first day. Tree combine-harvesting machines have HIAB cranes which can build the stack, but the horseman must take advantage of the terrain. Drainage ditches can be temporarily filled with poles, and embankments can be used to build the stack by rolling the poles in from above, so long as the HIAB on the timber lorry can reach in and drag them out. If there is no ditch or embankment, and no space at the roadside, building a stack on flat ground needs skids, levers, plenty of muscle and a bit of clever technique. Joss Rawlings in Greystoke Forest had showed me how to build a stack using a snatch-block with an open side, which allowed the direction of pull to be changed while the horse kept walking in a straight line.

Building a stack at Red Bank, looking down the hill.
(I am standing on the stack, 3 feet above ground level.)

At High Close, Ginger and I built the biggest stack of all our time together, measuring over 10 feet high. This was only possible because the terrain was right, the poles were big but not too big, and Ginger knew that he had to stop the load within six inches of the start of the stack, where

I had placed a specially cut skid or slide. As each pole was lined up, I could lever the butt, or lift up the tip a few inches to get it onto the skid, and then as Ginger pulled forward on a word, the end of the pole would slide up the skid, slide up the heap at an angle, and then drop into place to make a tidy stack. By the time it reached 10 feet high, the angle back to the ground was so steep that I had to fit a belly-band to stop the traces from lifting over the horse's back. The system was smooth and satisfying, and Ali took a series of pictures to show how it was done.

Red Bank – bait time.

The view from Red Bank Wood was magnificent, in all directions. In spring the wood was carpeted with bluebells and stitchwort, and the delicate green tracery of larch framed the landscape. To the north was the great mass of Helvellyn rising above Dunmail Raise, north-west was the Lion and Lamb and Easedale Tarn, with Allen Bank nestling at the foot, and the smoking chimneys of Grasmere village beyond the frozen lake. North east was Penny Rock Wood and the Coffin

Road to Rydal; to the south, behind the majestic trees of the Red Bank arboretum, was the mighty Wetherlam, Swirl Howe and Coniston Old Man. Away in the distance was Grizedale Forest and Near Sawrey, where our cottage sat in the low hills by Esthwaite Water, and beyond that was Morecambe Bay. The horizon was immense. One fine day in the middle of winter, I was sitting on the south edge of the wood at the end of the shift, exhausted, cup of tea in hand, contemplating the stack of timber and drinking in the majestic landscape around Wetherlam. There was snow and ice on the ground and a pale sun setting, a moon rising, with perfectly still air and a column of smoke curling slowly up from a cottage fire below. Out of the silence I heard from miles away across the valley the call of a lost fox hound, sounding clearly in the stillness. It was one of the most beautiful, plaintive, mournful and evocative sounds I ever heard, in a fairy-tale landscape. The whole scene still comes to mind as representing the best of the life and work of the woods.

Soon the larch at Red Bank was finished, and we moved straight on to first thinnings of a 'nurse' crop of larch among the hardwoods at Deerbolts Wood, a few hundred yards away to the north, on the steep slopes above the shingle shoreline at Grasmere. Deerbolts was north facing, with no sun at all in winter, so was icy as well as steep and progress was slow and dangerous. First thinnings are usually heavy work for the horseman, and I persuaded the National Trust that I needed help to gather the load among the standing trees on the steep hillside. They agreed, and Watto came to work with me, and he brought his own horse, Danny. I did not mention Watto to the Trust as he was on their black-list for some past misdemeanor, which neither they nor he would discuss, although I heard that it was something to do with barley wine. We managed to avoid injuries with two horses working on the icy slopes, but we were glad when it was done, even if it was the end of our occupation of High Close stables, and we moved the whole circus back to Ginger's stable at Harrowslack.

When spring came we moved our base once again, to Nibthwaite, at the southern end of Lake Coniston. Nibthwaite was a long pull from Sawrey, but as soon as the grass started to grow I found good grazing there, courtesy of the National Trust who owned the surrounding farmland, and I could leave Ginger to fend for himself at weekends. The photographs and my invoice book show that Nibthwaite was very productive, thanks to good terrain, a simple specification, good larch trees, and a good team.

At Nibthwaite we went back on to piecework, as I calculated I could do better than the hourly rate. I was extracting on good level ground with Ginger, Alistair Mawson was converting, and Watto was general roustabout, doing the measuring, clearing brash, helping with the stacking, sorting out the horse feed, and keeping us all entertained. Dallas Machell, the master wood cutter, was felling.

At Nibthwaite with Watto and Alistair Mawson

One of the many good turns Watto did for me was to introduce me to Dallas, who became my number one woodcutter. Dallas Machell was the son of Major Machell of Penny Bridge Hall, whose family had been at Penny Bridge since about the year 1200. When I first met him I assumed that the name Dallas referred to the city in Texas, but in fact he was descended from Rt.. Hon. Sir Robert Dallas, Lord Chief Justice, and he was heir to extensive lands and houses all around Hawkshead, Cartmel, Claife, Ulverston, and Dalton. Dallas had been sent to Eton College and although a wealthy man, he had little interest in anything except wood-cutting and trail hounds. As a woodman cutting for horse extraction he was in a league of his own. Not only could he drop each tree with an accuracy that was astonishing, but he would dress them out immaculately and at speed, turning each tree fully so that there were no branches or spikes left underneath to dig into the ground. He would methodically throw the cut brash out of the way as he worked, making a long tidy row so that neither the tree itself nor the extraction route was obstructed. Not only could you see the cut poles for the whole of their length, but the stocks were kept low, so the poles would glide out cleanly with no snags or tangles. Having Dallas cut wood was worth an extra man on the gang, and I made sure I paid him every Friday without fail, in cash. Perhaps I need not have been so scrupulous - each week he would take his pay-packet and stuff it into the glove compartment of his car, along

Nibthwaite: Fence posts at the roadside,
sawlogs and rustic poles behind.

with all the previous weeks' packets, unopened and still full of cash. He simply did not need the money. Dallas seldom went out except to walk his hounds, preferring to watch television, and he never took a holiday. So why did he work so hard and so carefully? I guessed that it was his pride, and a sense of tradition, and maybe a kind of distaste for his own leisured and moneyed class. He was independent, always his own man, and he developed unequalled skill as a woodcutter partly to earn respect and to show the world that he was not a toff, and partly for its own sake – the mark of a true craftsman.

Dallas was taciturn and stubborn. A Forestry Commission officer came into the wood where we were working at Nibthwaite to try to make Dallas wear the regulation chainsaw safety gear – gloves, helmet and safety boots. Dallas refused, and so the officer came to me, the main contractor, to tell me that he could not allow Dallas to carry on working. I called him over, told him what had been said, and asked him what he thought we should do. Dallas did not look at the officer, he just looked me in the eye and said "I think the best thing to do is to throw this fellow in the lake and not let him get out again." The Commission man retreated, muttering, and Dallas went back to work – in his wellies.

With Watto at Nibthwaite. (Bill wearing a 'brat' or leather apron.)

Not long after I finished at Nibthwaite I heard that Dallas had been killed in a motor accident while he was out walking his trail hounds on the highway at night. It was a real loss to me – he was the living proof that there are more important things in life than materialism, and that life can be given meaning by mastery of a practical skill. At the time I envied him his contempt for money, but I could not fault his attitude. Only the truly wealthy can live their life as they want, and to find a wealthy man choosing the hard life of the woods before the comfort of his satin cushions and fine wines in the big house sustained our idealism about what was important in life, even though it did not pay our bills.

Through Dallas and Watto, I got to know trail hounds and fox hounds, which were a vital part of traditional Westmorland life at that time. The hunt was still vigorous, and the hunt suppers and 'Merry Neets' were mighty entertainment. The love of the hunt was one of the few things that would ignite the passions of the dourest of Westmorland farmers, and each year the best singers would compete for a silver trophy, and I would travel miles to listen to them sing. At Yew Tree Tarn, in the late autumn of 1978, while we were working with a gang of National Trust woodcutters, the Coniston foxhounds came though in full cry, and one woodcutter, John Richardson from Langdale, could not resist the excitement. John was unusual in that he did not use a chainsaw, but still worked with the axe and the billhook, possibly because he did not have his safety tickets, but probably because he preferred it. As the fox ran past the tarn with the hounds 200 yards behind, John threw down his billhook and took off, whooping and yelling encouragement, and we did not see him again for three days.

The morality of the hunt was always a difficult and unresolved question for me. I did not follow the hounds, although I played and sang occasionally at local hunt suppers, where I found the men and women of the Lakeland foot packs to be mostly people after my own heart. The hunting ban made it illegal deliberately to hunt a fox with dogs, but for the great majority who hunted, the pleasure was not in the killing, nor in inflicting of cruelty and suffering. The killing had to be done anyway, one way or another, to control the foxes, and the ban did not stop the foxes from killing lambs and wiping out sheds full of chickens. The destructive cruelty of the fox was justified because it was labelled 'natural' whereas the hunters were considered unnatural and 'sadistic'. The Lakeland packs were not mounted, they hunted on foot, so only the fittest followers could keep up with the hounds, and most followers never even saw a kill. The pleasure of hunting on foot was a form of primitive ecstatic release, combining physical exertion with the skill and craft of the huntsman, the communality of the hunt followers, and the emotional high of the 'crack', the laughter, and the songs and tunes in the bar at the end of the day.

The Lakeland foot-packs which I observed had nothing to do with sadism, except perhaps for an occasional nasty brute of a type who exist in most cultures. It cannot be denied that the kill was the objective, but the suffering of the fox was not a source of pleasure. Our ancestors the apes have been predatory hunters throughout millions of years of evolutionary history, and it always struck me as a form of ethnocentric arrogance to assume that our human species would, or could, or should, deny that simple truth. There are over 270 species of carnivorous mammals which hunt, all of which cause suffering or 'seriously compromise the welfare' of their prey, and yet most of which are now protected. I have spoken to people who supported the hunting ban who would happily watch a nature programme showing a lion catching and killing a zebra, for their own recreation, and they would instantly send for the rat catcher to eradicate mother and baby rats. Is the rat catcher supposed to be full of guilt and remorse, or are they allowed to take any pride or satisfaction from their work? There are those who would not and could not kill a rat themselves, yet would happily make criminals of Westmorland farmers.

I would not now campaign for the hunting ban to be repealed – the page has turned, and our collective morality and democratic process has determined that hunting with dogs is unacceptable, but the ban showed how popular democratic opinion can be intolerant of the dissenting minority. It is easy enough to tolerate what you already approve of, so that hardly even counts as tolerance. Tolerating something you dislike is much harder. The hunting ban had no effect on me personally, and I did not march either in favour or against the ban, but I was uneasy at the way the ideology of the city took the heart out of a rural culture on the grounds of moral superiority.

'Merry Neet' in Far Sawrey

18. Loweswater
– fire eating and a feather bed

On the day the last wagon-load of larch was taken from Nibthwaite and I presented my invoice to Ken Parker, he asked me to move back to Loweswater in West Cumbria. I needed the work, but it meant a lot more travelling, and the distance was too great to commute. Ali now had a wee babe to look after, so I wanted to be at home as much as I could, but I agreed to work four days a week, with three days at home. The National Trust did not want me living in a caravan at Holme Wood, which was a favourite tourist beauty spot, and Bed and Breakfast was costly and so out of the question, so I camped in the bothy in Holme Wood during the week, and travelled home at weekends. Jim Chamley at Watergate Farm agreed to feed my horses while I was away.

Staying in the bothy was mildly romantic, but solo romantic does not really work for long, and turns soon to loneliness. The bothy was not exactly comfortable, especially after I injured myself again. I cannot now recall how exactly it happened, but a pole under tension was released suddenly, sprung back and laid open my shin, and the wound became infected. Damaged shins were an occupational hazard, and although I wore footballer's shin pads, I still have the scars from countless abrasions.

I took a few days off work and a few antibiotics helped me to recover, and when I went back to work, Mr. & Mrs. Chamley, out of kindness and sympathy offered me a room at Watergate Farm. I accepted without a second thought. Mr. Chamley was an ancient spindly farmer with the whitest of white hair and whiskers, and the twinkliest eyes I ever saw. I was convinced that he lived so long because he loved his life so much. He was kind and generous and entertaining, though his sense of humour was dry as only a Westmorland farmer can be. Barbara, Mrs. Chamley, was similar, but less dry, and maybe 20 years younger. They gave me breakfast and dinner and a proper packed lunch each day, and best of all I slept in their spare room, which was furnished with a 17th century oak four-poster and a proper feather bed.

Holme Wood Bothy, Loweswater

The true feather bed has vanished from our culture, which is a great loss to the world. Compared to the foam mattress, which is full of toxic chemicals and fossil fuels, the feather bed is a model of sustainable furnishing. It is also the most comfortable sleeping arrangement ever devised. It is like sleeping on a cloud. I cannot say if it is good for the back or bad for the back, but I can say that after working as a horse-logger all day long I would prefer to sleep on a feather bed with a feather pillow and feather quilt before anything you would find in a five-star hotel.

My second stint at Loweswater was memorable for yet another mishap, this one minor and self-inflicted. I had been fascinated by street entertainers ever since getting to know the great Johnny Eagle, the gypsy strongman, and I decided learn some of his tricks. During the long evenings after work, and in the hour-long lunch break when Ginger was feeding and resting, I had time to practise. My first choice of trick was learning how to crack a 20 foot tether chain, for which the first step was learning how to crack a stock-whip. (My great-great-uncle, Alexander Lloyd, was an early pioneer horseman in Queensland, and in his diary he described the unmistakable sound of a stock-whip, so I wanted to try it and find out for myself.) I made up a long whip from plaited sisal baler-band and practised every day until I could make it sound like a rifle shot, then practised some more until I could take a tin can off a fence post at about 15 feet.

Then I moved on to fire eating, with couple of tricks from a book called *Memoirs of a Sword Swallower* by Dan Mannix. I started with the 'Clouds of Smoke' trick. If you master this you can walk into a bar or some other public place, and suddenly, without warning, and with no cigarette or other visible cause, produce continuous thick clouds of white smoke from your mouth. The trick is to hollow out a walnut, drill a hole in each end, fill it with a mixture of sawdust and cotton wool soaked in saltpetre and then dry it out. When ready to perform the trick, you secretly light the cotton-wool so that it glows and smoulders but does not flame, and then you stick the two halves together with quick setting glue, and put it in your mouth. The theory is that the layer of saliva will form an insulating barrier to protect your mouth from burning, long enough to establish yourself as an innocent bystander, for maybe 5 or 10 minutes. You can then blow a stream of air through the walnut by means of the two holes drilled previously, and produce unexpected clouds of white smoke. I tried it twice, burned my mouth both times so I could hardly eat Barbara Chamley's delicious supper, and decided to try something else.

The next attempt was the standard but classic 'extinguishing a flaming torch by putting it in your mouth and closing your lips over it' trick. I had detailed instructions, and followed them exactly, but I still burned and blistered my mouth, even more badly, and this time I had to explain to the lovely Barbara Chamley why I could not eat her supper. After that I went back to the stock-whip. No stamina I suppose. I told my dad Walter about it and he quoted Lawrence of Arabia, who had said 'The trick is not to notice when it hurts.' I tried to explain to Walter that Lawrence of Arabia was reputed to have sado-masochistic inclinations, but that cut no ice at all. I never did make it as a fire-eater, and the burns healed and were soon forgotten.

Loweswater and the feather bed was soon just another sweet memory. Our daughter Eleanor was walking, summer had arrived, and I had more work to do at Harrowslack. We now had a proper signboard over the stable – 'Forest Horsepower' painted by father Walter's second wife, the multi-talented Gill Barron. The day I fixed it in place, I set Eleanor in front of me on Ginger's back, and she grasped the lead-rein with a tiny hand.

The new sign board – painted by Gill Barron

With the signboard came a proper letterhead, using a fine sepia image of Ginger pulling logs in Greystoke forest, painted by an accomplished artist friend, Stephen Darbishire. I proudly sent marketing letters to all the woodland owners I could find, extolling the advantages of selective thinning with horses, and although we certainly needed more work, really I was writing to show off the new letterhead.

Unfortunately the spring of 1982 saw the deployment of the tractor-mounted Norse winch from the National Trust wood yard at Boon Crag, Coniston, and my work for the Trust began to look a little uncertain. They still had no Norse winch in West Cumbria, so my next job was thinning another National Trust plantation at Isel, on the River Derwent, a few miles east of Cockermouth. I was allowed to have the caravan in the wood, so I towed it to Isel, the farmer dragged it into the wood, and Ginger and I carried on working away from home four nights a week.

The National Trust employed a different gang of woodcutters at Isel, and the main man in the cutting gang also worked at Sellafield, or Windscale as it had been before 1981. I quizzed him carefully about the plant. Ali and I had long been anti-nuclear campaigners, informed and encouraged by our friend Edward Acland, whom we had first met on a train full of people going to Trafalgar Square to protest against the proposed Magnox reprocessing facility in West Cumbria. Edward tirelessly campaigned against the proposed expansion, and he spoke eloquently and with passion. We agreed that the risks of using nuclear power to enable rampant consumerism were just too great, and soon formed an 'affinity group' of like minds to talk eco-politics and non-violent direct action.

At Isel I was in the company of a man who had worked at Sellafield for years, so in our tea breaks we naturally talked about the ecology and the politics of the Magnox plant. I asked him what he thought about the radiation emissions and the cluster of leukaemia deaths around Sellafield. Did he think the plant was dangerous? "I am sure it's dangerous" he replied. "But how do you know?" I asked. "I know because they pay double the rate inside the fence than they do outside the fence. They wouldn't do that if it wasn't dangerous!" I found the logic hard to fault.

Also at Isel I met a professional salmon poacher who came to look for me in the wood, wanting to buy firewood. We got to know each other, and before long he wanted to swap firewood for dodgy salmon. He fished the Derwent regularly with a wet suit and nets, wading up through each pool, working upstream and laying his catch in the bushes at the roadside. At the end of the night he and a mate would work back along the road in a van, picking up the fish and taking them straight to London. I am told that most of the salmon are now gone from the Derwent – probably due to his successful techniques. He confided his methods just before we finished the felling job, and he told me he had retired from poaching. Although the river bailiffs knew all about him, they hadn't been able catch him, so I am happy enough to reveal his secret 35 years later.

Isel had one major spin-off. I had time on my hands, so I expanded my tin-whistle repertoire. I had been playing on and off for years and in even 1973 I could earn more in an hour playing the tin whistle in Leicester Square than I could earn in a week working as a stage manager in the theatre, but I knew only a handful of tunes. At Isel I was alone in the caravan, night after night for weeks on end through the long dark nights, and although I had my radio, my tape recorder and a sackful of books, I would spend an hour or two each evening learning new tunes on the whistle.

Our social life centred on the music, which also provided some much needed income. After three years I had mastered the horse job, but we were still living hand-to-mouth, and my account book shows just how precarious were our finances. I struggled to earn enough with a horse to pay the bills at the end of the month. Too often things would seem be going well, only to have the van break down, or the chainsaw need a new chain, or I would bust a horse collar beyond repair. One wet day at Isel I managed to earn £22.00, against the odds, and as I was leaving the wood I spiked one of my safety wellies on a root and put a hole in it. The cost of a new pair of safety wellies was just £22.00. The words 'sick' and 'pig' came to mind.

When the horse work was scarce, it was the music that kept me busy and helped pay the bills. Ali and I had started a ceilidh band – *Devil's Gallop*. She played the fiddle and I was the dance caller, which I loved but meant that I missed out on the hours of solid practice that the band would get from playing gigs every week. I made up for it in the long lonely evenings in the caravan at Isel, where I learned a whole book full of Irish whistle tunes and a dozen or more songs. Those songs and tunes have earned me thousands of pounds in the 35 years since then. With the ceilidh band we could earn a double wage, and I also played a regular session with Steve Grundy in *The Cox'n's Cabin*, which brought in a few pounds more. On Sunday afternoons we played for beer and sandwiches at *The Hole In't Wall* in Bowness. Funds were tight, but life was sweet.

When Isel was clear, I towed the caravan on its last journey, to the Middlewood Permaculture Centre at Wray, near Lancaster, where it ended its days as accommodation for visitors, and I brought Ginger home. I had grown to love the magnificent landscape in West Cumbria, the smoky bothy, the feather bed, and the long dawn drive out to the West, past Thirlmere and Helvellyn and Castlerigg Stone Circle. Even so I was glad to be back home, sleeping in my own bed, with Ginger in his stable, watching our family grow.

Part of Devil's Gallop Ceilidh Band on the Lake Windermere steamer. Fiona on guitar, Ali on fiddle and Steve on melodeon. Marian Lloyd dancing, Eleanor behind.

19. Wigton - more horses & a taboo

Watto was still working with me as often as he could, and in my quest for more wages I was keen to try working two horses together, but finding an extra horse that could do the job, at a price I could afford, was not so easy. Good timber horses are not common. The work demands a different set of skills from the plough or the dray, and quiet temperament and sure-footedness are the vital qualities, no matter what the breed or what the size. A timber horse must be steady as a rock when working on steep slopes, often with little room to manoeuvre among many obstructions, the noise and fumes of chainsaws and heavy lorries at close quarters, and the swish of falling trees. I once saw Joss Rawling's old horse Charlie hit by the tip of a falling tree, and about 6 feet of it broke off over his back. Luckily it hit his saddle, and he hardly even flinched, but he was lucky. The longer I worked with Ginger, the more I realised how fortunate I was to find him, and how hard it would be to replace him.

An Ardennes mare would have been ideal, but at that time we could not afford one. (This was before the days when forestry horsemen can get a state grant to buy a horse.) The Ardennes is an exceptional breed, perfectly suited for heavy draught work. Strongly built with a high body weight but a low centre of gravity, they are extraordinarily quiet and even-tempered – placid to the point of stupidity. I had worked with the Ardennes horses at Geoff Morton's farm because the Morton boys preferred their Shire horses. The Ardennes had very consistent conformation and temperament and I was curious to know more about the methods the French used for breeding and breaking. I was told by Charlie Pinney, who imported Ardennes mares into this country, that this consistency came from selective breeding, not the breaking and schooling. I have never verified this, but he told me that under the French system, every pedigree Ardennes horse must be registered with the breed Society, and it is the Society officers, not the breeders, who decide which foals will be retained for breeding. Any animal which does not make the grade is killed for meat, with no appeal, so any horse showing bad temperament or bad conformation would simply be culled. It sounds severe, but it enabled the Ardennes breed to develop quickly, over a few generations, into a perfect draught horse. I found them to be mildly dull-witted, but they have the power and traction needed for working in the wood.

My (highly theoretical) business plan when I started out aimed to produce enough profit to buy two or three Ardennes mares, which would not only do all my work but also breed my replacements. It never happened and with hindsight it was never realistic, but it was a great regret that I could never afford to buy even one of these fine creatures. Thirty years later, the Ardennes horse has an assured place in horse-logging throughout the UK. In the Lake District, Simon Lenihan of Celtic Horse Logging runs a successful operation using Ardennes horses, although George Read, the senior working horseman in Cumbria, still uses his cobs and Galloways, which cost less to buy and less to keep.

I needed a second steady and experienced horse, and since an Ardennes was beyond our means, I asked Watto's advice. He knew of nothing suitable, but suggested that I call the main horse dealer in Cumbria, Ronnie Mowbray, so I called him, but he had nothing. As a long shot, I called Joe Huddart, the general dealer from Cockermouth. By coincidence Joe had decided to sell Sally, the Irish Draught mare I had borrowed when Ginger was sick with the strangles, and Joe was planning to take her to the

Wigton Horse Sales, a few weeks away. I talked it over with Watto, and we decided to go to the auction, and see what else was on offer. If nothing else was suitable, we would put in a bid for Sally. On the appointed day, Ali and I picked up Watto from his bungalow, and off we went to Wigton Horse Sales.

High Cross with John Sanderson

The annual Wigton Horse Sale was then the biggest horse auction in the North of England, and ran over two days. The October sale would usually be busy, with hundreds of horses of every breed, type and colour. There would be a few magnificent Clydesdales, with their manes and tails plaited and decorated, hooves oiled, and with gleaming white halters. In the next pen would be heavy hunters, next to a few strong cobs and piebald gypsy horses with so much feather their legs looked like hairy bell-bottom trousers. There were thoroughbreds, driving horses, hacking ponies, Shetlands, and dozens of Fell pony 'stags' - young colts, wild and unbroken, and taken off the fell just a day or two before, with a pen full of cuddies (donkeys) and countless rows of harness and horsey accessories. While we were inspecting the pens we met Ronnie Mowbray himself, an old friend of Watto. They shook hands, grinned at each other, and headed for the bar. It seemed that there was hardly a horse at the auction that had not been through Ronnie Mowbray's hands at some time, and he knew everyone. If Watto ever needed a snigging horse or a pony for his trekking business, he would call Ronnie Mowbray and tell him what he wanted. Invariably one would be found, because Ronnie was dealing with dozens of horses each week. He would try his best to match the buyer and seller, and had a reputation as a fair and honest man, who would even take back a horse if it turned out to be unsuitable.

At Wigton, Ronnie would buy horses for resale at a profit, and he valued his riding, driving, hunting and trekking customers because they were his bread and butter, but the 'base price' at an auction would always be the 'knacker price' or meat price. That price varies, but it is essential information when you are buying or selling a horse, as you can be sure that it would be unusual for a horse to sell at auction (or even through a dealer) for less than meat price.

I would use £1.00 per kilo as a rough rule of thumb. A 13 hand Fell Pony might weigh (approximately) 350 kilos. Ginger, a 16 hand heavy draught horse, weighed around 700 kilos. I was looking for a 500 kilo cob suitable for snigging, so if there were no other bidders except the butcher, I would expect to pay £500. If a horse did not make the meat price in the ring, the owner would not sell, and it would be withdrawn, although it might go to the meat man in a private deal at the end of the auction. The actual numbers and prices may be different 35 years later, but it is always a good idea to know the current meat price before going to an auction. How you find out is for you to discover!

It is a curious fact that many British and Irish people who will quite happily eat beef or mutton are physically disgusted by the idea of horsemeat, and would be nauseated if it was served to them on a plate. In Europe (particularly in France) it is considered no different from other meats, and it can be bought in butchers' shops and restaurants. Japan imports over 2,000 tons of horsemeat annually for making high value *sushi*, and for rendering into oil for cosmetics. Various explanations have been suggested as to why horsemeat *sushi* should be considered barbaric to British and Irish sensibilities, but no-one knows for sure. As far as I can tell it is an ancient cultural preference, and may be related to the notion of 'taboo'. That theory proposes that as recently as 8,000 years ago (a blink in evolutionary time) each Mesolithic tribe would have a different 'taboo' animal, which they would not hunt, thereby protecting that species with religious or sacred reverence while at the same time preserving it to be hunted by other tribes. The taboo culture persisted into the Neolithic period, when the land was settled by farmers, and from there it survived the Bronze Age and passed into modern times.

The taboo theory is one explanation for the fact that the Celtic people had a horse goddess, Epona, the only one of the Celtic gods or goddesses to be adopted by the Romans. Epona was widely revered by the Roman cavalry, many of whom were Celts, including both the Imperial Horse Guard and the Auxiliaries. They carried with them, across the Roman Empire, images of their horse goddess in miniature clay and bronze tokens, or else carved in stone and set up as altars. Images of Epona would be built into their stables and decorated with roses, presumably as protection for the cavalry horses. Two thousand years later, the horse still stands at the head of the animal hierarchy of human sympathy, probably higher even than cats. The British public still reserve a very special venom for anyone thought to be cruel to a horse.

The English Gypsies have retained their deep cultural affection for the horse, so that for many Gypsy families it remains a defining part of their identity. Although Gypsies are routinely stereotyped as cruel by the prejudiced popular press, the contrary is more likely to be true in my experience. Many Gypsies are fine horsemen, and for a few hundred years, until the motor-car came along, they were welcomed as good judges of horses, and welcomed as suppliers of the best and fastest animals. Before motor-cars, the man with the fastest horse was king, and for the gentry and aristocracy the fastest horse in the district was not just a practical and useful accessory, but a major status symbol.

The British distaste for horsemeat (especially in beefburgers) may have nothing to do with ancient Celtic or Roman reverence for horses – it may be simple sentimentality. After all we do not eat dogs, cats or rats, and as far as I know these were not Celtic goddesses, but whatever the origin of our reverence for the horse, it is deep seated.

Beef is now taken to mean exclusively 'dead cow,' but the word 'horse beef' has been and remains in common currency. The 2012 scandal, when unlabelled horsemeat was masquerading as beef, was a straightforward fraud in one sense – only a chemist could tell that the meat in a supermarket ready-meal came from a horse and not a cow, but it revealed a whole can of worms regarding the 'bottom end' of the meat trade. We had known for years that the inclusion of unspeakable parts of animals in pies and sausages was commonplace – an unsavoury development made possible by E-numbers and anti-oxidants, and made legal by Act of Parliament. When Ali and I sought refuge from Babylon and moved into the wood, we were converted to wholefoods and became very wary of any processed meat. For all we knew, the cans of minced beef were in fact cans of worms, and the bottom end was in fact what went in to donner kebabs. Once we had tasted our own home-made sausages, anything from a factory was clearly suspect.

What upset the leader writers about the horsemeat scandal was the fact that we had been deliberately deceived, but there remains more than just the whiff of a double standard about the hysteria. The verbal skills of advertising copywriters have been employed to mislead the public since literature was invented, and the description of horsemeat as 'beef' seems to me hardly worse than many other misleading adverts, but the idea of eating horsemeat touched a sensitive nerve.

Horsemeat is high in protein, low in fat and has a whole load of Omega 3s, but it remains deeply taboo in Great Britain, and equines have a special place in our consciousness, or more probably, in our subconsciousness, which recognises the nobility and sanctity of the horse. I experienced this when I tried eating horsemeat in a cafe, in Rochdale in 1977. It made me nauseous, and I could not eat it. I have never eaten it again – at least as far as I know.

Back at Wigton horse sales, after dragging Watto out of the bar and inspecting the horseflesh in the pens, we spotted a three-year-old Clydesdale gelding, and Joe Huddart's Irish Draught mare Sally. The Clydesdale was too young, too big, and unbroken but he was worth a low bid. We took our seats at the ringside, and waited for him to come through. We dropped out when the bidding reached £600, and he sold for over £1,000.00. We put in a bid for Joe Huddart's mare Sally and bought her for £510.00, with a good work collar thrown in, which was a fair price.

Coniston Water, East Side. European Larch poles

Sally was smaller than Ginger, about 15.2 hands, but strongly built, and she was in foal to a Clydesdale stallion, so we had a bonus when she produced a fine filly foal the next spring. We arranged for Ron Mowbray to transport her back to Harrowslack on Lake Windermere and we headed home. The return from Wigton that day was engraved for ever on our memories when the inebriated Watto, after drinking barley wine all day, was pleading to be let out of the car for a pee all the way from Keswick.

Eventually he demanded that we stop the car immediately, otherwise he would wet himself. We were in the middle of Ambleside, but we pulled over. Watto hopped out, and had a pee against the nearest wall. I will never forget Ali's wail: "Oh, no John! No! Not the Library!" We shuffled him back into the car as quick as we could, and got him home. Next day he remembered nothing, and had no hangover.

I yoked up Sally for the next job, among some tall European larch on the Eastern side of Claife Heights. I had agreed a piecework rate for the job, as I reckoned that everything was right to make money. I could work two horses together, which would, almost double the output in such good going. The terrain was perfect - a long, gentle downwards slope to the loading bay, and a good firm zig-zag track to get back to the top of the wood. The species was right - larch is the only deciduous conifer, and in winter it drops its needles, making for much lighter work and an airy wood. The side branches of larch are usually pulled off by friction as you drag out the poles, so there is less snedding, and what little there is can usually be done with a billhook, not a chainsaw. Because the side branches come off easily, the trees are quick and easy to fell, and hang-ups are rare. The wood was only 100 yards from my stable, and a short pull to the road, so there was very little time wasted in getting to work, and I could take the horses home for the midday feed. The job was to cut selective third thinnings, so they were big trees, which meant high value butts, and I could set a high price on each. I cannot now recall the price, but it was probably £2.00 a tree to fell and extract. The weather was perfect, and one day I pulled 80 trees with two horses. The felling was mostly done in advance, or while the horses were feeding - an hour or so in the morning and a couple of hours at midday. I made a note of my weights and measures and I averaged £70 per day for a week. Better.

Working two horses together is no problem as long as you have the right horses and the right terrain. Ginger knew his job long before he came to me, and because he knew his job he was more productive with voice commands than with any kind of reins or cords. After only a few weeks I did not bother with a bridle and bit, preferring to work him with voice commands in a head collar only. This offended the purists, and experienced horsemen would tell me that I was not in control without a bridle and bit.

I knew Ginger better than I knew my own children; I knew what he could do, and what he couldn't do, and I knew he could work safely to voice commands. Sometimes he could even work without voice commands, and if the terrain was right, (no stumps, snags or crooked trails) he could drag a load for a few hundred yards, stopping when he wanted, pulling on when he was ready and drawing up right beside the stack, inch perfect, without a word from me. He could not hook on without me, and he could not stack the poles without me at his head, but he could do just about everything else, with very little help.

First thinnings at Buttermere, Hooking off.

Sally, the Irish Draught mare, could do nothing on her own. She did not have the same experience as Ginger, and she needed to be worked with cords for a week or two. Once she settled down, she would follow him down the track and usually stop behind him when he stopped, provided she was not spooked. In the Harrowslack larch I would ride Sally (she was 4 inches shorter than Ginger so it was easier to get on her back without stirrups!) and lead Ginger along behind, up the zig-zag to the top of the wood. Ginger would be hooked on to a pole, and Sally would be hooked on to the one behind him, and they would follow the same track. Holding Sally's head, I would give Ginger the 'Gee up, get on' and away he would go. Sally would want to follow immediately, but she had to be held back until he was a good distance away, because she moved more quickly than he did, for fear of being left behind. Once Ginger had covered some ground and made 50 yards between them, I would drive Sally on down the same track.

Depending on the horse, it is generally safer to use voice commands, partly to keep a distance between the man and the horse, and also because of the space limitations in the wood. It is difficult to work long reins from directly behind the horse, as the load would be always around your ankles, so the best place to work with long reins is either beside the horse or beside the load. However, if you are beside the horse you often have to keep dodging round the standing trees all the way down. After a while you develop the skill of anticipating the standing tree that seems to be rushing towards you, then you quickly pass the reins in front of it with your right hand, and picking them up with your left hand while dodging round the tree to the left, so that you don't have to pass between the horse and the tree. If you get caught in a narrow gap between horse and tree, the horse will probably crush you. If it doesn't, then the load certainly will.

If I am making this sound highly dangerous, that is because it is highly dangerous, and looking back 30 years I am fairly sure that my addiction to the 'adrenalin rush' is bound up with my time in the woods. Which was cause and which was effect I cannot say, but I am sure I took some foolish risks, and I have no doubt that I was lucky to do seven years without serious injury. I came very close and had a couple of visits to hospital, to say nothing of permanently scarred shins. I am not dwelling on the dangers to show how brave I was (foolish would probably be a better description,) nor even to warn aspiring horse loggers not to do it. In fact I would recommend that anyone interested should try it, and they will find out soon enough not only how dangerous it might be, but also how to manage the risks. I had committed myself to a dangerous line of work, and once committed, I just had to get on with it.

I was well aware that if I worked with horses in the wood I risked losing a finger, or worse. Because I needed all my fingers for my musical activities, I always wore heavy gloves and tried to avoid the obvious dangers. Although working two horses at once was possible, it was high risk. If I had a clone of Ginger the Wonder Horse I could perhaps have managed two horses all the time, but Sally was a bit dim, barely fit, inexperienced and unpredictable. Even though I could pull and stack about 12 tons of larch on a good day with two horses going together, I preferred to work solo with my bomb-proof Ginger companion, simply because we understood each other.

Even though I had what might seem to an outsider to be a telepathic understanding with Ginger, and he (usually) obeyed my commands, I disliked it when journalists attributed human qualities to the horses. I was in no doubt that my relationship with Ginger developed and matured as we got to know each other, to the point where I felt that there was a two way flow of communication, but he was always a horse. I doubt he ever 'tried to tell me something' although I grew to recognise the meanings of certain behaviours. After a few months working with him I got to know the signs when he was weary, or hungry, or nervous, and even when he was in high spirits and wanted a good gallop. I knew that he would respond to my tone of voice if I wanted to calm him or gee him up, but apart from 'Gee up' and 'Whoa' I don't believe he knew the specific meanings of words - although he was probably capable of learning more if I had the time to teach him. It may be that I underestimated the effect of his dependence on me, and I certainly came to love him in the sense that I thought about him a lot, and I liked to be with him, but I did not treat him as if he was a human, and he did not treat me as if he had stepped out of a Disney film. He was a sentient creature, and intelligent as only a horse can be, but he was clearly a horse and not a human. Nevertheless, journalists regularly described him as if he was a fairy tale creature.

As a literary device, anthropomorphism is strongly associated with art and with storytelling, and it has ancient roots. I had no problem with anthropomorphism in children's stories, myths, fables, religions or fairy tales, but I objected when it was used to describe my horses, as it seemed to me to devalue our relationship. I thought it a lazy and sentimental literary device, designed to elicit lazy and sentimental responses. There was a steady steam of reporters and feature writers who came to watch us working, some good and some bad. Those that wrote about Ginger 'nodding his agreement', giving me 'questioning looks' or 'shaking his head in disapproval' missed the point. He responded to my voice, he quite liked music, and he just loved being scratched where it itched, but he was completely hopeless at poetry and not great at conversation.

Although Sally was dim by comparison and mildly unpredictable, and although she did not always understand '*Gee up;*' and '*Whoa*', she was mighty powerful. Although I preferred solo work with Ginger, if I gave Sally a straight pulling job, with no obstacles and good terrain, she was a match for him, as I found out when we took our next job, pulling a hay sled full of Christmas trees off the mountain side at Grasmere.

Westmorland Hay Sled, by Tinne
Reproduced by courtesy of Abbot Hall, Lakeland Arts Trust.

20. Newby Bridge - *horses vs. machines*

In October 1982, I had a call from the owner of a Norway spruce plantation on the steep hillside behind the Travellers Rest on the edge of Grasmere Village. He managed the plantation for Christmas trees, and wanted to try horse extraction to avoid damage to the roads and tracks caused by the forestry tractors dragging out loaded trailers and climbing back up the hill. The wheel ruts soon became rivers in heavy rain, and washed out the hardened track, which cost him thousands of pounds to repair. He thought the horse would do less damage, and he was right. I took the job, (piecework, at so much per tree.)

I reckoned that we needed sledges rather than carts so as to keep the centre of gravity low. The owner agreed to pay for the sledges, and I did the rounds of a few farm sales until I found two ancient horse-drawn hay-sledges, used for getting loose hay and bracken bedding off the steep hills into the hay barns of Garsdale. We trailered them to our stable yard, fettled them up with steel plates and new runners, and started hauling Christmas trees.

It was sheer slog, and in the end the pay was only average, but Sally saved the day. In fact I had made a bad bargain when setting my price, not realising how much profit there was in Christmas trees. The rate should have been much higher, particularly since we saved the owner all the damage from the tractor. According to my diary, we pulled out 3,480 running feet of Christmas tree in a typical day and a half, which he sold in Liverpool at £0.65 per foot, so he drew £2,240, of which he paid us £150.

The hardest part of the job was hauling the sledge back up the hill, and after the first trip Sally was not so keen to try it at her usual rush, so she settled in to a steady pull. We took a good rest at the top while the work gang loaded up trees, but the uphill pull was so strenuous that we made sure we pulled maximum loads going down again, so as to keep the number of trips to a minimum. The horses needed to rest on the descent, and although Sally had settled down, she was unfit, so the sweat poured off her in steaming streams. It looked worse than it was, but the boss's wife thought the horses were working too hard. I disagreed, politely, and I tried to explain that the horses were taking no harm, and pointed out that they only needed a word, not a whip, to get them moving, but it made no difference – we were paid off. The sledges were sadly left behind in the wood, presumably to rot, and we retreated to work on our home ground at Claife Heights until the new year.

The Christmas tree job soon had a spin-off. The wagon driver who hauled the Christmas trees to Liverpool worked for the Lowther Estate, He told his boss about me, and I had a call from the Lowther Forestry agent who had over a hundred tons of timber to fell and extract at Chapel House Planting, near Newby Bridge.

The Lowther Estate had their own woodland, but they also acted as merchants, buying standing wood and employing contractors to fell, extract and convert, and then selling it on to wherever they could get the best price. The job at Chapel House was third selective thinnings of mixed species on a steep but workable banking, with a long straight pull, through a wall gap and into a loading bay. The ground was too wet and steep for most tractors, and the pull was too long for most tractor winch cables, which are limited by the size of the winch drum. There was not enough volume to justify rigging a skyline system, which in any case is not really efficient for selective thinning, only for line thinning, so the contractors were not keen on the logistics. Because Lowther had called me, not the other way round, I reckoned that I had an edge. The contract was offered for tender – the lowest bid price would win. I bid for the work at a price per ton higher than a tractor man might expect, as I reckoned that there would be no tractor men bidding, otherwise Lowther would not have called me. As anyone used to competitive tendering will know, if you get the job you are probably too cheap, and if you don't get it you are probably too expensive, but at Chapel House Wood we wanted the job and we got it.

Negotiating prices was a vital part of my work. Most skills can be developed with practice, but to be a good negotiator you also need to be merciless. If you are too soft, you will be fleeced. I was much too soft when I started out, but I hardened up with a little practice. When paid by the ton (as opposed to an hourly or day rate,) I could choose to be paid either by stack measure, (converted to tons) or by a weighbridge ticket. The stack measure depended on the percentage of airspace and the notional number of tons per cubic metre, both of which were necessary to calculate the weight, and both had to be negotiated. A stack of conifer roundwood normally has about 35% airspace, but that might vary from 30% to 40% depending how well it is stacked and the straightness of the poles, and the final measure could only be agreed with the merchant after the wood had been stacked. The weight of a cubic metre of cut timber also varies with the species of tree. I could not control the figures used by the merchant for the weight per cubic metre or for the airspace in the stack, so I would calculate the numbers for myself. My result seldom tallied with theirs, but there was no way I could 'withdraw my labour' after the wood was stacked, and I quickly learned to agree the variables in advance.

Merchants never paid in advance, due to partly to the scallies and cowboys who would take the money and run and partly because wood paid for in advance had a curious habit of disappearing overnight. The weighbridge ticket was more reliable than the stack measure, but even that was uncertain. The ticket weight depended on the length of time a stack would be left sitting at the roadside, where it would dry out and lose weight through evaporation of the water content. We were living 'hand-to- mouth' and could not wait days or weeks for a weighbridge ticket to support my invoice, so there was always a trade-off.

Negotiation with merchants was tricky enough for a tractor man using a Norse winch, or for a hydraulic forwarder driver who never left his cab, but they could handle in an hour what took me a day, so they had a wider margin for error. For me the negotiation was often too difficult and too disheartening. I had to fell the trees by hand, dress them out by hand, turn them by hand, gather a load for the horse by lifting and dragging by my own muscle-power. I had to lift or roll the butts if they snagged on the way out, lift them again to stack the poles, and finally lift and carry each 2 or 3 metre piece when converting and stacking for the wagon. The bottom line was that I might have one ton of wood stacked at the roadside, but I had moved that same ton two or three times with my own hands at different points in the harvesting process. After all that, by the tricks and formulas proposed by the man who held the cheque book and all the cards, I might get paid for only three quarters of a ton.

My negotiation was weakened by the fact that in commercial forestry I was always up against the machines, and could not compete on piecework except in selective thinning when the machines could not get in to the wood. If the machines could not do it and we got the work, we could command a higher price per ton because we could not produce as much volume in a day. At Chapel House Wood the horses could get the wood to the roadside cheaper than the machines, so we got our higher price, but even with everything in our favour, it was barely enough. We were in third thinnings, the terrain and the specification were both good. I had two good horses, skilled woodcutters and a roustabout helper, so we were efficient and productive. The wagon came to collect once a week, and my invoices were paid on time. We were just about showing a small profit so long as nothing went wrong, although it was sheer slog.

We did however have good accommodation for the horses. Chapel House was on the east side of Windermere, about half a mile from the foot of the lake and about half an hour from home in Sawrey. Thanks to Watto I was introduced to Mrs. Fell, who had a good field nearby with a dilapidated old shelter and in lieu of a week's rental for grazing we spent half a day replacing planks on the field shelter and fixing the leaking roof.

I took the trouble to make a neat and tidy job, as Mrs. Fell was generous in letting us use her field, and that care and trouble proved the truth of the saying that 'what goes around comes around.' By another strange coincidence, I bought Mrs. Fell's field 20 years later, and my brother Tom now uses that same field shelter as a woodshed, so we ended up with most of the benefit of the repairs.

I found good stabling near the wood, in the old coach house at Chapel House, with the kind and generous Mrs. Bindloss. It was wintertime and there was no electricity, so Watto and I spent a day cleaning out yet another cluttered old Victorian stable and installing paraffin hurricane lamps, then carted a load of hay from Windermere. I signed up a team of woodcutters and we were ready to go.

Bankhead Cottage: Horses vs machines – commuting to work

I walked Sally and Ginger to Chapel House from Harrowslack, riding one and leading the other on another long and glorious trek through the winter woodlands down the west side of Windermere, and bedded them down on fresh straw in the Chapel House stable.

The days were short, so we started work early. After feeding horses and sharpening saws by the light of the paraffin lamps, we walked a mile up the forest roads through the snow and ice to the loading bay as the sun was rising. It was midwinter, so we had about five hours of work if we wanted to get home before dark.

I had two new woodcutters – John Sanderson, ex National Trust and now freelance, and his mate, whose name I did not record. John used the 'shortwood' system, in which the crosscutting is done in the wood as soon as the tree is felled. The chainsaw man fixes his tape measure to the butt, and crosscuts the pole as he works along the length, dressing out the side branches. When he makes the last crosscut he releases the tape with a flick, it winds back into the reel on a spring, and he moves on. The shortwood system works well for extraction by 'all terrain' forwarding machines, which pick up each piece with a hydraulic grab. Shortwood has advantages for the woodcutter, who does not have to dress out to the tip of the tree, which is usually wasted after the last crosscut is made at the stack, so the conversion to a saleable product is done as part of the dressing out operation, which saves time. Shortwood can be extracted with a horse, and it can be easier to make up a load from 3m lengths rather than long poles, although for very short pieces (fence post length and similar) a sledge is more efficient.

The wood cutters worked fast, and with help from George van Weinen working as assistant horseman and roustabout to gather the loads, we had a 20 ton load stacked at the roadside in record time. The wagon duly arrived, and I asked the driver where he was taking the load. The specification was for '3 metre pulp' so I expected him to say 'Bowaters.'

Bowater had a paper mill near Barrow-in-Furness where they made, among other things, Andrex toilet paper, but the driver wasn't going to Bowaters, at least not directly – he was going via Hull, 120 miles away, where the wood was to be unloaded on the dock side and shipped to a paper mill in Norway. His wagon would then load up on the dock side in Hull with paper pulp from the same Norwegian mill, and haul it back 120 miles to Barrow-in-Furness to make Andrex toilet tissue.

This was my first taste of globalisation. My ideal of using heavy horses instead of fossil-fuel powered machines to enable a balanced, holistic, organic, sustainable, deeply green future, independent of Babylon, was in the end supplying the global timber trade. I was working my backside off to earn a living by extracting pulp wood that would be transported 120 miles across the country and 500 miles across the North Sea, only to be transported back again to within 15 miles of where it started, and all to make toilet paper! Not ideal, not mythic, not heroic, and hardly radical environmental action. Was this really what I set out to do?

My ideology was being tested against the market, and taking some hits. Maybe the mountain was unclimbable after all. My heavy horses were up against heavy capital and heavy mechanisation. I could not compete in commercial forestry except in special circumstances where everything was in my favour, like Chapel House, but after I had done the deals, paid all the gang, fed the horses, and fixed the motors, I could hardly make a living wage and had become just another cog in the machine.

I could still make a living when being paid by the hour and not by weight, but the hourly pay depended on selective thinning, amenity woodland, conservation values or enlightened subsidy. Although those values were just around they corner, in 1982 they had not yet arrived. In commercial forestry, line thinning or no thinning at all was the dominant method of harvesting, and the woodland owners and merchants did not care how I got the wood out. I could have felled the trees with a stone axe and dragged them out with my bare hands for all they cared. They paid a price per ton, for a stack at the roadside, and they cared not a fig how it was produced.

As the tree combine-harvester machines became more efficient, the cost of producing timber was falling. I tried every way I could to work out a competitive price and still make the job pay, but even with two horses was struggling, and not just because I was too soft at negotiating. Even at maximum efficiency, I was at the mercy of the market, and with a small family to feed, the mission began to look hopeless. I could not expect muscle and bone to compete indefinitely against iron, steel and diesel oil, if, after a month of solid graft, I could not pay the bills.

There were still some sympathetic woodland owners, notably Ken Parker at the National Trust and Jon Williams at the Lake District

National Park. They were not driven by the need for profit, and they had thousands of acres of selective thinning and amenity woodland which was best worked with horses. As long as they were willing to pay an hourly rate, there was some hope for me, but Chapel House Wood demonstrated that even in good going we had little chance of competing in commercial forestry, even with two horses and an efficient gang. I could only survive in the long term in the subsidised world of amenity and conservation.

That was a wake-up, but I was not yet ready to throw in the towel. The underlying logic of our ideology had not changed, and the calls for more environmentally sensitive and sustainable economics were growing louder, so I decided instead to cut my costs and diversify. I advertised Sally and her lovely yearling Clydesdale-cross filly foal, Biddy, for sale in the *Farmers Guardian* and collected £1000.00 in crisp notes. We kept hold of Ginger, we kept our National Trust cottage and our pukka stables at Harrowslack, and raised my prices as much as I dared. We held as tight we could to our ideals of a sustainable way to make a living while the insanity raged around us. Reality was biting, so we branched out.

Timberjack Harvester (Copyright Heikke Valve)

21. Horseplay - *music, film and TV*

Diversification turned out to be not as difficult as I had imagined, and brought more adventures. Music was the obvious first step. Ali and I had been playing music together since leaving London in 1974, and several years of long dark winter evenings in a caravan in the middle of a forest up a mountain in West Cumbria had given me the opportunity to practise my instruments and learn enough material to make up a concert set of songs, tunes and stories. We played mostly Irish traditional music. Ali was a competent fiddle player and a patient teacher, and I played the banjo, the tin whistle and the piccolo, and had developed a singing style which could be heard un-amplified in a noisy bar. To feed our growing appetite for tunes we would travel to local Irish music sessions in pubs and private musical gatherings, and were drawn to the Lancaster 'Clock's Back' Festival and even as far as Girvan in South West Scotland

In 1981 we had formed our ceilidh band - *Devil's Gallop.* The name was taken from the name of the back road to Hawkshead by Esthwaite Water, near Sawrey - the site of mysterious goings-on which had passed into local legend. Ali played the fiddle, and I called the dances, with occasional bursts on the piccolo once the dancers were competent enough to be left un-guided for a minute or two. The band was comprised of local musicians we had met through the Irish sessions – Steve Grundy played the melodeon, Fiona Loynes on guitar and Liz Rowe on concertina. We were later joined by Malcolm Fielding on mandola/bouzouki, and we played dance music for ceilidhs, weddings and parties, We would take our infants along with us and put them to bed at the back of the stage, until they grew up enough to want to dance all night instead of sleeping.

The vicar of Hawkshead un-surprisingly disapproved of the name of the band, so we would get no bookings from the Church, but we soon found that we were in demand and could earn some extra cash, drink plenty of beer and dance into the night – a good combination, and highly recommended. At that time there seemed to be only three Ceilidh bands in Cumbria – the ground breaking *Ellen Valley Band* near Caldbeck, *Tumbling Tom* from Furness, and our own band, so before long we were busy. Our first paid engagement was the New Year's Ceilidh for the Brathay Expeditionary Group, and they had such a high time that we were immediately booked again for the following year.

Playing for a dance band is without doubt one of the quickest ways to master an instrument. The tunes are repeated over and over, so the twists of the melody and the underlying dance rhythm are soon learned, and the repetition imprints the tunes on the mind and the fingers, so that it becomes possible to play the tune without having to think about it. That frees the mind to focus on the nuances, the ornamentation, the key changes and the endings, and it is those details that mark out a good dance band from an average dance band. We were not the best, but we created a buzzing energy and an atmosphere that could carry a roomful of people into ecstatic dance territory. More than once we experienced that primitive, positive, semi-trance which is created by wild music and the willingness to throw yourself around the dance floor. It was there I discovered the truly communal nature of the dance. The most popular dances were those where intricate cooperation with your partner, with the other dancers in the set, and with the other sets on the floor, provided the dopamine, the adrenalin, the lift, the excitement, and the satisfaction.

'As long as you keep drinking, you can keep dancing, and as

long as you are dancing you can keep drinking'. *Strip the Willow, The Cumberland Square Eight,* the 'basket', the reels, and the Irish two-handed swing, combined with moderate intake of alcohol, became our recreation as well as a vital source of income.

Devil's Gallop Ceilidh Band, in Ambleside

The dance band provided entertainment, and a modest amount of cash, but we were still on the breadline. I looked around for another job that paid a wage and was flexible enough to fit in with the horse work. After the New Year's Eve dance for the Brathay Expeditionary Group, I had made contacts at Brathay Hall, and I applied for a job as an external tutor, taking groups on week-long adventure education courses to learn about team building and leadership skills. Despite my lack of qualifications, I got through the interview and another door opened.

The Brathay experience changed my life - again. All the tutors who worked there were first required to participate in a training course as students and my first course was a revelation. Brathay was at least as formative for me as the disciplines of drama school, the technical training of stage management, the creative intensity of the Arvon Foundation, and the demanding logistical and physical tests of horse-logging. The Brathay technique was to build a team rapidly, by supervised exposure to shared risks of various kinds. The participants learned immediately that they simply had to cooperate and trust each other. If you are dangling off a rope half way up a cliff for the first time in your life, and that life is in the hands of someone you only met that morning and who has never handled a rope before, you discover that trust can be a real and difficult choice, and that fear is a physical emotion, not just an abstract training concept. You learn that skill in handling a rope is not just a training objective, but a life or death responsibility. When every move in an elaborate training game is watched, noted down and then judged by a merciless observer in a candid review of your performance, you learn a lot about yourself and about what makes a good manager. Although the Brathay training had to be physically safe, with 'belt and braces' standards, it was often physically exposed and usually emotionally dangerous. On one Brathay course my group of foreign-currency traders from major banks was reduced to tears of frustration because of the mental stress of the game they were playing – and they did it to themselves. At Brathay, the tutor is both leader and observer, and the trick is to know when to intervene and when to step back and let things happen. I tutored all sorts of groups, ranging from bankers from the City of London, to vulnerable boys and girls in the care of a Local Authority.

Brathay Hall paid well, it was exciting, and I was able to combine tutoring with my horse work by taking groups on a horse-drawn expedition in a bow-top wagon pulled by Ginger round Coniston Water.. On one such adventure, a well-balanced but unfortunate young woman tried to kill a tutor by beating him over the head with a very big stick. He was a drama specialist, and had dressed up as an ogre in a cloak and a mask, carrying a flaming torch, and had surprised the group of young teenagers by appearing suddenly out of the woods while they were having supper round a camp fire at Parkamoor Barn. The group panicked and rushed inside, screaming with genuine terror, and slammed the door. The 'ogre' tried to force his way in, thinking the game was going well.

The unfortunate young woman's instincts were to protect her friends from this threat, so she quite reasonably tried to kill him by hitting him with a club. Luckily for both of them she did not succeed, but it was an extraordinary demonstration of the power of loyalty and team building , and the power of a woman's instinct to protect.

Brathay Hall was a financial life-saver, and unforgettable, but it was occasional work, and not a long term solution to our grinding financial problems, By another stroke of good fortune, the solution appeared courtesy of Beatrix Potter. Her legacy of woodland had provided my employment with the National Trust, and provided our fine cottage, and was about to provide our primary source of income. I began work as a supplier of horses and other animals for film and TV productions.

The BBC announced that they were to make a film for TV, provisionally titled *Tales of Beatrix Potter*, a dramatised version of the writer's life. Since we lived in Near Sawrey where she had lived and farmed, we soon heard on the grapevine that the BBC crew would be based in the village and were looking for horses and handlers. I called them up, winkled out the name of the Properties Buyer and invited him to visit us. He duly arrived at our meagre cottage in a huge car, dripping with gold bracelets and requests. By the time he left, and after enjoying Ali's special cream tea, he had ordered two heavy work horses, a driving horse and harness, two farm carts with harness, a trap, a small herd of native-breed cattle, an Aylesbury duck, twenty day-old ducklings, a donkey, a flock of geese, a herd of sheep, four goats, three dead crows, and a fox. Within a week he revised the order and added traditional milk churns, leather suitcases, station trolleys and a horse-drawn milk-float. I was offered a walk-on part as a the milkman to drive it. I put my timber-horse work on hold for a month to get the whole thing together and decided that maybe Babylon had been under-rated.

The sheep came in the horsebox from Rod Everett at Middlewood. The geese came from Mr. Wood the chemist in Staveley, sitting quietly with their heads sticking out of sacks, so that they could not run or fly. The day-old ducklings arrived by rail. We acquired a fox cub from a farmer in Langdale, and it lived in a laundry basket and then a rabbit hutch, becoming friendly and playful.

Once we arrived on the film set, the horses behaved perfectly, the fox cub and the ducklings could charm the hardest of hearts, the director was delighted, and we were invited to join him to drink champagne to celebrate the 100th shot. Things were looking up. By the time the film was done, I had turned over almost £2,000 for three weeks' work – rather more than I had earned in the whole of the previous year. Sure, I had some expenses, I worked exceptionally long hours and took some risks, but I knew which side my bread was buttered. Diversification was good.

With Eleanor, fox cub and ducklings at Bankhead, Near Sawrey

With a head start from the BBC, I was soon supplying horses and other animals, props and locations, to film companies. I had first done film work with Geoff Morton and his Shire horses, and it was obvious that film and TV work was safer, more exciting and better paid than hauling timber. I was still working with horses, but now I could mix the hard graft with the relatively easy money, and had a better chance of making a living.

Bill as a milkman in 'Tales of Beatrix Potter'

I called the new business 'Animal Connections' and advertised my services as widely as I could afford. Within a year or two I had worked as a horseman on *The Gathering Seed,* for the BBC, and on *Blackjack* for Ken Loach and Tony Garner. I hired Zebu cows from Lord McAlpine, (then treasurer of the Conservative Party) for the ITV series *The Jewel in The Crown*. I worked as a rat handler for the *Blackadder* series, and provided a Gypsy wagon for the Granada TV series *Sherlock Holmes*. I provided horses for a film, *Briggflatts,* about my poet hero Basil Bunting and found the locations for Beeban Kidron's first full length movie *Vroom.*

Film work with animals is surprisingly un-glamorous. Animal

handlers are usually called to be on set an hour before anyone else, as producers assume that something will go wrong and they need to be 100% sure that any animals will be there on time. Once the animals arrive on the set, it is not unusual to hang around for 4-6 hours before your shot is called, and then hang around some more waiting for the light to be right, or waiting for the aeroplane noise to go away, or waiting for the prima donna to turn up. The monotony would be broken by sudden bursts of activity between shots and by the meals. Excellent food was provided for the actors and crew by specialist film-set caterers, and I soon learned that the trick was to get a 'crew' meal-ticket and not an 'extras' meal ticket!

Filming *Briggflatts – The Basil Bunting Story* in Garsdale

I did not make my fortune in TV work, and soon realised that once again I could not compete with the heavy money in the film business – I was just too small. I had almost no financial reserves, whereas most suppliers of film props and animals were heavily capitalised, with large warehouses full of props and whole farms full of animals of every kind.

Filming Ken Loach's *Blackjack*, with Joyce Smith (Mrs. Carmody)
Bob Pegg and Tony Garnett in the background

The full story of *Animal Connections* is too long to tell in a book about horse-logging, but the following anecdotes will give a flavour. One morning I had a call from the props buyer at the BBC, who by then was a regular contact, to ask if I could supply a dozen black rats for the next episode of *Blackadder*, Rowan Atkinson and Tony Robinson's comedy TV series. Although I had never handled rats, I knew that the correct reply is always 'Yes, no problem', to take the job and then sort it out. The rats had to be on collars and leads, so would have to be trained and then delivered to the location at Alnwick Castle on a date about three months away. I did a hasty calculation (or guess) as to how long the training might take, estimated the travel and accommodation costs, and decided a suitable fee. We agreed a price (high), they sent me a contract, and I started looking around for black rats.

Black rats are *Rattus Rattus*, whereas brown rats (the common kind) are *Rattus Norvegicus*. Black rats are believed to have carried the plague (which is why Rowan Atkinson wanted them for a gag about 'lucky black rats') so they are usually exterminated wherever they are found. This could be a tricky job.

I telephoned the rat catchers at all the main sea ports. (I had been in the Merchant Navy, and knew that all sea ports had rats.) I telephoned the right people, who of course denied vehemently that they had any rats at all. Next I tried my local rat catcher, who advised me very strongly not to take the job. He wore very thick protective clothing when handling rats, as a bite could pass on a very nasty disease, and he scared me off with tales of lingering fatal illness. Still no rats. Hmm. A week had now gone by, and I was getting worried.

At that time I was attending a course in small business planning, staying in the Midland Hotel in Morecambe. While walking along the prom at lunchtime in search of a potato pie, I spotted a pet shop and in the window was a family of white rats – with black heads! I asked inside, and the owner told me that they were domestic rats, (i.e. pet rats, *Rattus Rattus,*) related to brown rats but without the plague bit, and bred in various combinations of black and white. I bought one, a female, as a trial, then called in at Woolworths for some black *Grecian 2000* hair dye, and nipped back to the hotel, where I dyed the rest of ratty's body with a toothbrush. I locked her in the bathroom to dry off while I spent the afternoon struggling with cash flow forecasting in the seminar room, and hurried back to my room at 5.00 p.m. to see how she looked. Perfect job

– shiny black hair all over, no blotches, no join, and no roots showing. I named her Rati, which is 'beautiful lady' in Indian.

I made up a lead from my dressing gown cord, and experimented by taking Rati for a walk around the bedroom. Perfect again – she would move forward on the lead quite happily, and allowed herself to be handled, so I reckoned I was in business. Next morning I put her in a cardboard box marked DO NOT TOUCH, left the box in my wardrobe, and went down to breakfast and the morning business session. At lunchtime I checked back, only to find that she had chewed through a corner of the cardboard and escaped. A quick search revealed her sitting happily in the corner, so I just about got away with it, although Hotel Reception were very frosty. The chambermaid had spotted Rati while cleaning the room and had a fright, so I was told to take her away. (Luckily she had a silk ribbon round her neck, otherwise the maid might not have known that she was a pet and sent for the exterminator.)

I just had time to dash back to the pet shop, where I bought the whole family – 14 rats with a cage thrown in, all for £70.00 – and skipped off the rest of the day's coursework to drive home to Sawrey. (I decided that was probably better for my cash flow than another afternoon of book-work.) I spent the rest of the day dyeing rats in the bath, and by bed-time I had them all looking like film-stars. For the next few weeks I would handle them for half an hour every day, and after a few days they would stay on my shoulders while I wandered around the house. One or two of them even seemed to like banjo music. I made up choker leads using leather boot laces and fuse wire, fitted carefully round their necks. They would travel well enough along the ground on leads, except that in a bunch they would get tangled up very easily, and the problem of untangling a dozen rats still comes to me in dreams occasionally – shades of Orwell's Room 101.

One day I left them tied up by their boot lace leads in a row along the wall, and after half an hour, one rat bit through the leather lead and hid under an armchair. Within half an hour most of them had done the same, which led me to conclude that rats are careful observers and learn fast. They were easy to handle, very intelligent, and much less troublesome than most animals I handled.

The day came for the filming, and I set off in the orange BMW with the cage of rats under a blanket in the boot. I stayed overnight with friends in Wensleydale, and because the BBC wanted me on the set at 8.00 a.m., I set off from Hawes at 5.30 a.m. for the long drive to Alnwick. As soon as I arrived I was sent to 'costume and make up', where I was dressed in a monkish brown robe, medieval hat, pointy shoes, a grotesque false nose, greasy wig, and a lot of facial warts. Pleased with my new *persona* I collected up the rats, and put them up the voluminous sleeves of my costume. They would run up and down the sleeves, peeping out of my collar, and sit on my belly, wriggling under the robe. Weird, and mildly disturbing, but funny and strangely fascinating. Looking like something out of a space horror movie I was driven to the set, where I entertained the extras and crew with rat tricks while waiting for Blackadder and Baldrick to arrive in their Winnebago. We had developed a clever routine in which four rats, in line abreast, would jump from one forearm to the other forearm, and while I rotated my arms parallel to each other in front of my chest, it looked as if the rats were galloping along, which in fact they were. Once Blackadder and Baldrick arrived there was further delay while they had their breakfast and discussed the day's shooting. When they emerged, I was told that they had re-written the script, and the rats' scene had been written out.

I was mighty disappointed not to show off my skills as a rat trainer on national TV, but as a consolation for being written out, I kept my bit part, standing in the street as a rat seller. The gag was that Blackadder thought that rats would ward off the plague, so I was to stand shouting "Get your lucky black rats here!" I appeared for about a quarter of a second in the final edit, but because my bit-part entitled me to be a 'contracted extra' rather than a mere animal supplier, I collected repeat fees, and for 20 years afterwards I would occasionally receive small cheques from the BBC in respect of videotapes sold in Mexico or New Zealand, or obscure broadcasts on UK cable TV. I came home only slightly disappointed that I had not become an overnight 'novelty act' TV sensation, and then had to decide what to do with the rats. Half of them went to The Queen

Katherine School in Kendal, where they lived out their days in the biology department, and the other half went to sit on the counter of a punk record shop in Carlisle owned by one of the crazies at the Seahorse (Midland) Hotel where I had first acquired them. He wanted them 'to attract the customers'.

My next show-business job came when I had a call from John Fox who asked me to provide a heavy draught horse for the Ulverston Lantern Parade. The event was staged by street-theatre designers *Welfare State International, the Engineers of the Imagination,* founded by John Fox and Sue Gill. I had first worked with them in Burnley back in 1976, and our paths had crossed occasionally.

We boxed Ginger to Conishead Priory, home of the Manjushri Institute, a Buddhist monastery, where the Welfare State makers were building huge puppets and lanterns for the procession. Ginger was to carry on his back a model ship almost as big as himself, and inside the ship was a large sculpture of a tree, embellished with lit paper lanterns.

The whole caboodle was about eight feet tall before it even went on his back, and I did not like the look of it. I gave Ginger a pair of tight fitting blinkers so he could not see his extraordinary load, and the makers attached some very dodgy bits of string which held it all in place. My script required me to make an entrance, leading Ginger, complete with ship and tree, into the middle of the huge crowd. This would happen at the end of the main action, and represented the tree of life floating off into the night just before the finale.

The tree was wobbly, the boat was precarious and the girths were barely adequate. I expressed serious doubts about the rig, but was persuaded that my role was crucial, so I agreed. What Mr. Fox had failed to tell me was that Ginger's arrival was the signal for fireworks. And not just any fireworks, but the detonation of the largest mortar shell that I had ever heard. It was totally unexpected, and the noise of the initial explosion was like a blow in the stomach. I felt the ground shake. I had a very tight grip on Ginger's bridle, but I felt him go rigid in a spasm of fear.

I cannot prove it with a picture, but I was sure at the time that he raised all four feet off the ground at the same time. Luckily for me, and luckily for Welfare State's insurers, he did not bolt, the tree stayed in the ship and the ship stayed on his back, but it was a very close call. I was probably more scared than he was, and we left in a hurry, vowing never to work for Welfare State again. As far as I know they were not even aware that it might be a problem, which just goes to show how an understanding of working horses has vanished from everyday consciousness. We patched it up with Welfare State, but that was the last time I joined the Ulverston Lantern Parade.

Animal Connections: Ulverston Lantern Parade
with Welfare State International

My next film job was finding locations for Beeban Kidron's first movie *Vroom* – the story of two boys from Blackburn who go on the run in a restored vintage American car and hide out in an isolated farmhouse in the Lake District. I had taken an advert in *Spotlight*, the Film and TV casting and locations handbook, and I had a call from Beeban wanting a location finder and fixer in the Lake District. I was hired, and for several days I hunted round the hills for a place which would meet the specification. Eventually I thought I had found it - a wonderful old farmhouse in an isolated valley, complete with broken stone cornices, Georgian panes in the windows, cobwebs dangling from the beams, and a photogenic landscape. In great excitement I called Beeban, who immediately sent her designer First Class from London to check it out. We drove up into the hills, and walked the last mile up a broken track to the farmhouse. As soon as she saw the yard she loved it, and the inside of the house was even better - lime-washed walls, cast iron range, stone tables in the larder, all the works - tatty but very chic. Up the creaky oak staircase were three bedrooms along a dark corridor. She opened the door of the end bedroom, and horror of horrors - chalked on the inside of the door in capital letters it said "EXTRAS, COSTUME & MAKE-UP".

"What's this?" she said, sharply "Has this been used before?" Well, obviously it had, but I really, truly, honestly had no idea - it was as much a surprise to me as it was to her. She was convinced I was trying to sell her a used location - definitely 'damaged goods' in the film business - and we drove back to Shap in silence for me to call the owner from a phone box. "Has that farmhouse at Wet Sleddale been used before?" "Oh, yes, a few times" "What for?" "Well the last time it was a film called *Withnail and I.*"

I had never heard of *Withnail and I*, as it was not yet released, but my suspicious designer friend knew all about it. The location I had found was Sleddale Hall, better known as Uncle Monty's Cottage in the cult film, and after shooting the scenes the film company had not bothered restoring it to the state they found it. The picturesque stone cornices were actually made of polystyrene, and the cobwebs came out of a spray can. Eventually I convinced her that I was innocent of deception, which she must have accepted, as she asked me to keep on looking and went back to London. Eventually I found her the right place - Parkamoor House, above Coniston Water, and she was able to do a deal with the National Trust. Later I heard that the Yankee car had to be lifted in by helicopter, but that might be folklore. The film *Vroom* was finished, and I believe it was screened at the Cannes festival, but I don't think it ever made it as far as General Release in the UK, so I never saw it until years later when it was screened on TV, and thirty years after that managed to buy an old video tape.

As part of the diversification strategy, I tried operating horse-drawn rides for tourists on the Glebe in Bowness. I had identified a possible route, along a road that was closed to motor vehicles, ideal for horse-drawn rides, and right next to the coach park so busy with visitors. I bought a large flat cart, once used for hauling hay, from a farm sale at Godmond Hall, near Burneside, and drove it home about 12 miles on a glorious summer day, standing up like a charioteer, singing 'Joe the Carrier Lad' all the way home. Next day I fitted it with seats and painted it red.

The flat cart at Tinkler Wood, minus the seats.

Although it was a two-wheeled flat cart, it was designed as a farm hay cart, and needed a different set of harness from the flat carts intended for driving on the road. Driving gears use leather tugs or heavy leather loops to support the shafts, and the britching is fixed with a leather strap to a metal eye on the shafts. The leather traces run from the collar right back to a hook at the far end of the shafts. Cart gears are different: the shafts have a fitting with three hooks, one to hook on the short chain traces, one to hook on the britching and one to fix a chain, which passes over the horses back, through a metal channel on a specially constructed saddle, and down to hook on the other shaft. I knew what was needed, as all Morton's horses used cart gears, but I did not have a set. Providence intervened, and within a fortnight another farm sale was advertised near Hawkshead, and I purchased a fabulous set of cart gears, heavily inlaid with brass. I polished up the brass work so that Ginger looked like something out of a Hollywood epic, groomed him like he had never been groomed before, plaited his mane and tail, polished his hooves, and set bells on his harness. I painted an advertising banner to hang off the side of the cart, then crossed over the ferry to try and earn some tourist dollars.

The police were not impressed. Two constables approached looking stern. They told me I needed a licence. I told them I had checked with the local authority and that I did not need a licence. They told me that the road was closed. I pointed out that a Traffic Regulation Order did not apply to horses. Not wanting to be outsmarted, they said that if I did not stop they would arrest me anyway and find something to charge me with when we got to the station. I stopped.

22. Woodmanship - new directions

Between the film work, Brathay Hall adventures, lantern processions and the thwarted pleasure-rides, I was still looking for horse extraction work, but after laying off the gang and selling Sally and Biddy, I was working solo, felling my own wood and extracting with just one horse. I was 'on the books' with Lowther Forestry, and when a contract came up to work selective thinnings in Grizedale Forest they invited me to tender for a job not far from home at Tinkler Wood, just off the road from Coniston to Hawkshead. The terrain looked good and the cutting specification was simple enough - fence posts, pulp and sawlogs, with a good bonus in the form of rustic poles.

The market for rustic poles was too small for Lowthers to bother with and they were not usually included in the specification. Starting from the butt end of the tree, sawlogs would be crosscut at the point where the diameter of the pole was 9" (under-bark measurement), regardless of the length, and the next section, down to 5" top diameter, was crosscut for wood pulp in 3m lengths. Finally fence posts would be cut at 5' 6" long, with a top diameter of 3" minimum.. Conifers in commercial plantations usually grow straight all the way to the tip of the tree, but anything less than 3" diameter (i.e. the top of the tree) would normally be counted as waste. Woodcutters working for machines would cross-cut fenceposts down to 3" diameter and not bother dressing out the waste, which would be left lying in the wood. I knew that I could dress the top 10 feet of each tree, down to 1" diameter, and take a rustic pole out of every one. After the logs were dragged to the production bay and converted, I would be left with a 10 foot rustic pole, 3" at the butt and 1" top diameter - ideal for gardens. At the end of the week I would load up the stack of rustics and deliver them to garden centres all over the north of England, where I could sell them by the hundred at £1.20 each. That would make an extra two days' wages per week, for little extra work.

After Chapel House Wood, I decided not to do piecework unless I could earn as much as my day rate and I was wary of tendering too low, so I added 50% to my usual piecework rate, and slightly to my surprise, got the job. Lowther agreed that I could take the rustic poles for myself, and they brought me a good bonus for a few months, but that was too good to last. The garden centres found that they could buy rustic poles in container loads all the way from China, at half the price I charged. Globalisation had hit me again, but at least I had a good piecework rate.

I had been in Tinkler Wood for a week when I had a call from Tony Jackson, an aspiring forestry horseman from Ivybridge in Devon. He had a bought a strong cob work horse, Beauty, and had read about my horse-logging activities in a draught-horse magazine. He wanted to learn about snigging, and asked if I would take him on as an apprentice. I agreed, offering him the same terms that I had started out on - he would be paid what he earned, no more and no less. I told him that he could start as soon as he liked, but he would have to find his own accommodation. He replied that he did not need accommodation and would camp in the wood, which I did not take seriously until he did it.

Tony packed up his kit in home-made saddlebags and set off to walk from Ivybridge to the Lake District. He arrived about three weeks later and knocked on our cottage door. I gave him a cup of tea and took him out to Tinkler Wood to look at the job. Within a day or two he had made himself a rough camp shelter out of ropes and canvas slung between the trees, bodged up a set of snigging gears from old car seat belts, and started pulling wood. He was keen to learn, Beauty was a powerful horse, and before long he was earning a meagre living.

Beauty and Ginger seemed to get along, although they distracted each other so much that working them together could be complicated. Wherever one horse was, there the other wanted to be. When Beauty was loaded up and set off down the track, Ginger wanted to follow, whether I was ready to go or not. Because I worked to voice commands, if Ginger chose to ignore my shout and set off before he was hooked on, there was nothing I could do except follow him down the hill, empty handed, cursing him and Tony and Beauty, and trudge back up again to pick up the load. The horses' attachment to each other encouraged a distinct reluctance to work, and they became more or less inseparable. One day I turned up for work to find them gone, not standing waiting for their food as usual. We followed their tracks for miles, wandering the maze of forest roads in Grizedale, sometimes retracing their steps in great loops. Generally they were heading south, and eventually we found them five miles away, at the most extreme southerly point of the forest, their chests up against a high stone wall, gazing wistfully at a field full of fresh new grass.

Two horses and two horsemen in the same wood made for too much friction. Apart from managing the wandering horses, my main problem was that since laying off the gang I now had no woodcutters, so I had to fell and convert enough wood to keep Tony and Beauty going, and I had less time for snigging myself. Tony did not have a chainsaw and he did not want one, and so I had little option but to cut for him as well as for myself. .Although I could charge him for my chainsaw work on his wood, I was a horseman not a chainsaw man, and meanwhile my horse was standing idle. It could not last long.

After a few weeks Tony was confident enough to go on his own, and he moved to Middlewood Permaculture Centre, near Wray, to extract firewood. He stayed there a while and then we lost touch, but I will never forget him, not least because wild garlic seemed to be the single most regular part of his diet. He was an eccentric, but there was no malice in him, and we became friends. He had a genuine love for horses and children, and he did much good charitable and educational work on his smallholding in Devon. He was one of the few people who shared with me a similar blind faith about working horses, and was motivated by the same spirit that had prompted us to start horse-logging. He left Middlewood to go back to Devon and about a year later he send me a few copies of his children's book *Beauty's Journey to Lakeland,* written about his road trip to work with Mr. Lloyd in the woods in the Lake District.

Tinkler Wood – Stacking European larch for fencing strainers

While working in Tinkler Wood, in September 1982, our second lovely daughter, Marian, was born. A week or so later, I was woken up on Sunday morning, after a late night out playing the banjo, by a phone call from a farmer neighbour who needed 200 fence posts in a hurry. I rushed off to Grizedale to cut his posts. Partly because I was hung-over, partly because I was in a hurry, and partly because I was careless, I managed to put the chainsaw into my foot. Luckily I was wearing steel toecaps, but the bar bounced off the toecap and cut into my instep between the toecap and the Kevlar protection in my trousers, which only reached down to my ankles. I could have timed it better, as Ali could not leave the infant Marian, so I had to call on a neighbour to me into Kendal to get stitched up yet again.

While the wound was healing, a National Trust job appeared unexpectedly. They needed a horse to pull 150 poles off some inaccessible rocks in Kentmere so that they could get a winch to them. I walked Ginger through Ambleside, up the Troutbeck valley, over the Garburn Road and into Kentmere. Watto and I had a laid-back couple of days pulling Scots Pine before walking home again, back to Tinkler Wood, where my stack of wood had been drying out for weeks while I was injured. Half of what was left had been stolen, along with my bolster sledge, which was a sentimental loss as I had made it myself five years before.

The job at Tinkler Wood took us until the end of the year, and I took another temporary job at Brathay Hall to get us through the winter. In the spring of 1983 the price of converted timber at the roadside began to fall, and by Christmas that year it was down to £10.00 per ton. The economy was just beginning to lift out of recession, but the manufacturing base was giving way to service industries, unemployment was pushing 3 million, inflation was around 10%, and the mining industry, which used thousands of tons of home-grown timber as pit props, was on its knees. £10 per ton at the roadside was a lower price than when I started out seven years earlier, and meanwhile inflation had driven up my costs by about 60%. At £10 per ton, I would be working at a loss. My base price was too high, so there was no work. My expenditure now exceeded my income, and for the first time we had to claim Family Income Supplement.

The dole kept us afloat for a few weeks, but there was no point in keeping on the horse business unless we could make it pay. Luckily, my work as a management trainer at Brathay Hall had brought me into contact with business analysts, personnel officers and high flying managers, and it was clear that our business model needed some serious review. One result of this review was the formation in 1983 of Woodmanship Ltd, which combined my experience of traditional woodland skills and my involvement with the Brathay Woodland Project. I got to know Michael Gee (Amenity and Conservation Planner and Landscape specialist at

Old Brathay) and Tom Clare (County Archaeologist), and together we established Woodmanship Limited, a registered charity trading as The New Woodmanship Trust. We met once a week for months to formulate our plans, consult with lawyers, hammer out a constitution and obtain charitable status.

Our first project was to invite Oliver Rackham, OBE, to speak on *Landscape and British Woodlands* at the Theatre in the Forest at Grizedale. The event sold out, and demonstrated that we had an audience. We formed a plan to acquire the old ironworks at Backbarrow, at that time a derelict site, but which had once been the biggest user of charcoal in the whole of England. We planned to take a lease on the site, restore the furnaces, restock the vast charcoal stores from local woodlands, and open a National Museum of Woodmanship, thereby at a stroke re-invigorating the whole of the Furness coppice industry.

It was a bold, ambitious, imaginative project. We were confident, and wrong. We surveyed the site, drew up plans, successfully negotiated with the National Park planners for outline planning consent. We invited heavy friends to come on board, and we gathered support from everywhere we could, but serious money was, as always, an insurmountable problem. We had support in principle and a small grant from the Dartington Amenity Research Trust which we used to commission a wonderful scale model of our proposed Backbarrow National Woodmanship Centre, designed by a student of John Makepeace the furniture designer. In the end we ran out of time, ran out of steam, and eventually the owners called a halt and looked for buyers elsewhere.

As part of our fund-raising plan we decided to raise our profile and invite subscriptions from a national membership, which we would launch at a *Weekend in the Woods.* Six months before, as part of my diversification plan I had conceived the first Lake District *Weekend in the Woods*, which was scheduled to take place at Cat Crag, a fine old house with a patch of woodland above the Western shore of Lake Windermere. Cat Crag belonged to entrepreneur Paul Hughes, who was sympathetic to our ambitions, and I booked the house and grounds as the venue for the event. I produced a brochure, advertising a weekend course in felling, horse-extraction and timber conversion, swill baskets, charcoal, turnery and besom making. It was all set up and ready to go, but two weeks before the event it was clear that there were not enough takers who had paid their fees in advance to guarantee the tutors' wages, so I had to cancel. With the New Woodmanship Trust and the help of Michael Gee and Tom Clare, we arranged a bigger and better *Weekend in the Woods* at Brantwood, Ruskin's home on Coniston Water, and launched our national membership campaign.

The second attempt was a success, thanks to Michael Gee's contacts, foresight, and his skill at marketing. Using the basic Cat Crag structure we added guided walks, curragh building, iron smelting, chair bodging, pole lathes, a craft tent and a steam powered bobbin lathe. We advertised widely, brought in two marquees and high class caterers, and the people flocked in. After the weekend was over we had 1,800 people through the gate, a profit of £700.00 in the bank, and 200 paid-up members, all shareholders in the New Woodmanship Trust. We were up and rolling.

The *Weekend in the Woods* convinced us that we were on to something. We had identified an appetite from the general public for

Horse extraction on Nab Scar, by Tinne
Reproduced by courtesy of Abbot Hall, Lakeland Arts Trust.

traditional woodland skills, woodland experiences and education, craft skills and woodland management for amenity and conservation, a world away from the mechanised dark plantations of commercial Sitka spruce. I spent most of my spare time for the next six months with Michael and Tom devising a training scheme for woodland apprentices. This was to be 'the big one,' that would establish the New Woodmanship Trust as a major player. At a time of high unemployment and anxiety about the environment, it was a winner and we knew it. The new direction in woodland management would need skilled people and we planned to offer the first national training course.

Participants would learn all the skills necessary to make a living as woodmen and women, covering marketing, accounting, health and safety, estimating, job costing and negotiating as well as the practical skills of chainsaw use, silviculture and coppice management. There would be options for learning swill basket making, besoms, hurdles, and wood-turning. The apprentices would get a year's financial support in the form of a training grant from the Manpower Services Commission, or MSC, which was the job-creation quango of the day

Naturally I wrote myself into the project proposal as Project Manager, on a decent salary, plus expenses, although the job would be a publicly advertised appointment and there was no guarantee that I would even get an interview. As I had nothing in the way of formal qualifications, a degree in Drama and Theatre Arts would count for less than nothing when applying for a job as a Woodland Skills Training Manager. Nevertheless I was sure there would be a job for me somewhere in the project, and I was keen to make it happen.

To improve my chances, I enrolled on a two week course for small business start-ups, offered by the same Manpower Services Commission that we hoped would finance our project. The course was run at the Midland Hotel in Morecambe, (the scene of the *Blackadder* black rat episode) - an icon of art-deco architecture, then in its 'late seedy' phase. We all wore shirts and ties, and sat in rows learning how to write a business plan, read a balance sheet and develop a marketing segmentation analysis. It was a new world for me, working nine-to-five, wearing a tie, and keeping my pencil sharp, but I soaked it up. In particular, it gave me the skills to write a detailed business plan and budget for the apprenticeship scheme, and for the first time I was able to prepare my own annual accounts for the tax man. I drafted the hands-on woodland skills part of the syllabus, and added a few pages of business planning and management training. Michael Gee wrote the education and the conservation sections, and typed up the proposal to submit to the MSC. We had support from the wildcard Woodland Officer from the Brathay Woodland Project, Peter Coates, who wrote the silvicultural section.

Peter was a wildcard because while he was working on the Brathay Woodlands project, he was also hatching a very elaborate scheme to take over the island of Jamaica. This would be achieved by staging a military *coup d'etat,* led by him, and which he was convinced was supported by Her Majesty the Queen. He tried his best to involve me in his plan, but I kept well out of it, saying that I was too busy. I tried to pop his balloon before he got himself shot, but he was impervious, almost obsessed. An unlikely story, I know, but true. He was a kind and funny man, and became a good friend. As far as I know Jamaica is still an independent constitutional monarchy. You meet some extraordinary people in the wood.

Soon we had a prospectus for training people in a radical new form of woodland management, with a focus on amenity, conservation, craft products, coppice and charcoal. My diary entry reads *'We need a new labour-force of environmentally conscious, semi-entrepreneurial woodmen with new skills.'* We believed that the thousands of acres of derelict coppice woods in Furness was the right place do it and that we would attract people from all over the country. When the prospectus was complete we made our application for formal approval from the Manpower Services Commission, to provide the finance to make it possible.

The MSC liked the look of it, gave it the initial thumbs up and put it through their system for vetting by other government agencies. Everyone liked it, except the Forestry Commission, who vetoed it on the grounds that it was unnecessary. *'There is no market for small scale coppice wood in the South Lakes, and therefore no need for a training scheme'* they said.

At the time they were right in one way, but only because we were ahead of the wave and they had not caught up. We could see the potential in neglected woodland, and our 'Weekend in the Woods' was empirical evidence of serious public demand, but I could not deny that I was unable to make a living as a horseman in competition with globalisation, mass markets and heavy machinery. We failed to justify an expensive scheme to train more people so that they too could fail to make a living, and it was little comfort that, three years later, the Forestry Commission did a graceful about-turn and marched in the opposite direction.

We thought their rejection was wrong and short-sighted. I went to see John Voysey, then Forest Manager at Grizedale and known to be sympathetic to our case. I gave him all the evidence of a groundswell of change from *The Weekend in the Woods*, the letters of support, and the credentials of our heavy friends. He was kind enough to listen, and he even took me on a trip to speak at a conference on woodland conservation, but he also played Devil's Advocate, pointing out that the Forestry Commission was established to build up a strategic reserve of timber - and they wanted big timber, not scrubby underwood, coppice products and amenity woods full of wildlife (or pests as he called them). He recognised the public appetite for amenity and conservation, but he had no power to change the decision which represented the entrenched traditional view of the Forestry Commission. We thought they were wrong, but that was the end of that. A year's work wasted.

We were proved to be ahead of the times, but by then it was too late. We should perhaps have persevered and tried again, but the momentum had gone. By 1987, three years later, John Voysey himself spoke at a meeting of Nature Conservancy Council staff in Grizedale on the subject of conservation in Forestry Commission plantations. That meeting announced a major strategic change in Forestry Commission Policy.[1] Conservation, which until then had been seen as a constraint on their timber growing activities, was now embraced as a core objective. That change resonated throughout the whole of the woodland and forestry economy, and before long central government was falling over

---

1 *Woodland Conservation and Management* By G.F. Peterke

itself to give grants for coppice management, broadleaf planting, selective thinning, diversification of *flora* and *fauna*, amenity woodland, nature trails, and even conservation training.

The timing of that shift in policy was one of the great disappointments of my life. I had predicted the change seven years earlier, prepared for it, advocated it, demonstrated it, bet heavily on it, and went it alone. I got there early, turned my back while waiting for the boat to come in, and when I looked again the boat and come and gone without me. I felt as if I had bought a winning lottery ticket and the dog had eaten it. Fifteen years later, the Bill Hogarth Memorial Apprenticeship Trust (BHMAT) and the Woodland Pioneer courses in South Lakeland began doing almost exactly what we proposed in 1983. It was a hard lesson, but there are no prizes for being ahead of the times, just disappointment and frustration. Maybe the Forestry Commission was bull-headed and short sighted, or maybe we were too ambitious, or maybe not ambitious enough, or a dozen other reasons, but the fact was we could not get it together. That proposal was the high point of my time in the woods, but when it failed, I was back struggling to feed a family and a heavy draught horse. The price of timber was still falling and I lost heart.

Luckily for my self-esteem, it is never over until it is over, and the next chapter in our woodland adventure, Project Phoenix, was a winner. After the MSC Woodland Training Scheme was turned down, we applied to Jon Williams Head Forester at the Lake District National Park (or the Lake District Special Planning Board as it was then called) to support a pilot scheme for the re-establishment of commercial charcoal burning in the Lake District. The revived charcoal industry would provide local employment and stimulate the revitalization of thousands of acres of broadleaved coppice woodland in Furness, with real benefits for amenity, species diversity and the woodland economy. Michael Gee in particular deserves the credit for his conviction that the scheme could work and for his determination to get the funds.

With help from Michael and Tom Clare, I drafted a production and marketing plan for an application for pilot funding. Michael sharpened it up, added chapters to cover the landscape, conservation, amenity and tourist dimensions, and processed the proposal on his all singing, all dancing, Amstrad Word Processor. (We were amazed to see a dot matrix printer producing page after page of text, printing in two directions without human intervention.) The application for funds was submitted, and was successful. A grant of £1,500 was approved and Project Phoenix was launched.

Now we were moving. We allowed twelve months' lead time to establish a suitable woodland site, arrange for charcoal kilns to be made to our specification, and to find, cut, stack and season a good supply of oak and alder charcoal wood. Michael and Tom were both in full-time jobs, so I planned to take on the job of managing the project, producing the wood, doing the trial burning and the marketing, and writing the report.

We had a fair wind in our sails at last, but before the 12 months was up, my old workhorse Ginger died. Just as we embarked on our most successful project, my mainstay was gone, and our woodland dream ended suddenly. I woke up in another world. So it goes. Flow my tears.

Project Phoenix test burns: Walter's camp at Thwaite Head
Photo Gill Barron

Lakeland Charcoal at Finsthwaite

200 bags of Lakeland Charcoal, ready for delivery.

23. *La Partenza* - *the parting*

In December 1983 my faithful Ginger horse developed liver failure, or maybe cancer, and had to be destroyed. He was stabled at Pull Wyke when he went off his feed and lost weight fast, his eyes went yellow, and his urine smelled vile. The vet pursed his lips and shook his head. It was one of the hardest days of my life, although I have had worse since. Ronnie Mowbray took him, and he probably went for meat, or maybe to feed the lions at the zoo. I have often, many times, wondered about the morality of my sending him to the knacker after he had served me so well and taught me so much, and more than anything else it was that decision that broke the spell of my obsession with work horses.

I had few very options about how Ginger and I should part. I went to see Watto and asked him what he would do, trying not to let my emotion show. "Call Ronnie Mowbray" he said, as I knew he would.

I have never owned a pet dog, although I had a working sheepdog when I was a child, but I have seen how people find it very hard to replace their dog after it has died – it seems disloyal somehow, an empty gesture that brings no comfort, just a constant reminder of the loss. Dog owners can at least have their pet quietly put to sleep at the vet's surgery, maybe holding its head or its paw, and if they wish they can bury the body in their garden or under a tree somewhere, with a sapling to mark it. I have done that with a foal, more than once, but not a 700 kilo draught horse.

I could have walked him into the knacker wagon and killed him there, but I had seen that done too and the process is not pretty – a bolt gun, a sharp knife, and lots of blood. I would have had to bring in a slaughter man, since I did not have the skill or the courage to do it myself, and slaughter men don't like a confined space, so he would have wanted to kill him on the ground, then winch his body in to the wagon and winch him out again at the other end. Brutal, emotionally harrowing, and very expensive – the whole thing would have cost more money than I had in the bank, even if I wanted to do it. Not really an option.

The Ronnie Mowbray way was better. In fact it was the only option, so I took it. Ginger would walk into the wagon, he would be on his feet when he left me, just as he was when he found me. Watto telephoned Ronnie for me. He drove into the yard the next morning in his familiar wagon, turned it around, cracked a joke, looked at Ginger, and offered me a price. I didn't argue. I led Ginger in like a lamb, tied his halter, gazed into his yellowing eyes, rubbed his nose, stroked his neck, and said my goodbye. I shut the tail ramp myself, in silence. Ronnie gently thrust a wad of money into my pocket, not my hand. He pulled away in the wagon, and seven years of my life rode out with him. Good years, great years, probably my best years.

When you work with livestock, you have to live with dead stock – so it goes. I had seen death on the farm since the day I could walk, so that was nothing new, but the loss of my horse was a different matter. We had a rapport; he knew my moods and I knew his; I looked after him, he made my living; we worked hard, we had fun. A man and draught horse can have real fun, believe me, especially on a warm spring morning galloping around a great big field under a blue sky or a golden moon.

One of the hardest things for me that day and in the few days that followed was keeping up the macho front with Watto and the older men who had seen it all before. Though they might have a sentimental streak, they never, ever, let it show – not for livestock, although they might weep at a funeral, and Watto would weep over nothing when he was drunk. When I told him how it had gone and my voice cracked a little, the look on his face said "Life and death, boy. Live with it." What he actually said was "Here, have a drink." The macho front was not merely to preserve a tough image in public, as is often thought, but simply to prevent the self from breaking down and causing more pain. I wept for a few minutes in the hay barn, and that evening at Bankhead I poured a libation for the old horse as I watched the sun go down over Wetherlam.

I had learned at University that the ancient Greeks used libations to discharge strong emotion safely, and I have done it twice in my life. It is a simple procedure, but powerful and effective, and it surprises me that the practice has completely died out of our culture. All it takes is a natural supply of running water and a bucket or a large bowl. The water is scooped out of the pool with the bowl, and using both hands it is slowly and repetitively poured into the earth. The extended and repeated act of scooping and pouring water provides a particular emotional release that is hard to find in any other way. After a few minutes of repeated movement, the body and the mind enter a different state, which one might describe as trance-like. With that state comes fluidity, clarity, a loss of self, and a sense of cleansing, of purity. One can surrender to the emotion without self pity, one can weep without shame, and when it is over, one feels released. For the Greeks it was a regular religious practice, and demonstrated humility and piety. For me it was a simple discharge of tension, especially after a death or loss. It may bring the tears, or it may stop the tears, but however it works, the repetitive stooping, straightening, and the slow and steady pouring can ease the mind and the heart.

That libation for Ginger helped me overcome the first tug of the peculiar guilt that has never really left me since. The guilt was not because could I not save him - I had rescued him from the knacker before he came to me, and he had a good seven years of bonus time - and not because I had failed to take enough care of him - I did everything possible to keep him healthy and never begrudged any cost. So why should I feel guilty? I knew that the only way for him to go was the Ronnie Mowbray way, and my guilt was unreasonable, and in a way sentimental. Anyway, maybe it wasn't guilt at all but grief? How could I know? How could anyone know? After all these years, I think I felt bad just because I could not find a decent, better way for him to go. I did not know exactly how he would spend his last hours, so I felt that I had betrayed him. I guess I just wanted to be with him when he died, but that too is a sentimental indulgence. I sometimes reflect on what else I could have done, and a few scenarios come to mind, but they are all fantasies. It was just hard to say goodbye, that's all.

There is an ancient, persistent, brutal notion in many cultures which in Western pop culture can be summed us as 'there is no such thing as a free lunch.' The idea lies at the root of sacrificial rituals. If things are going well, especially when they are going really, really well, when the atmosphere at a party just inexplicably lifts off, for example, or great good fortune comes your way and stays around for a while, or a touch of fairy dust settles on your hair one day and you feel blessed - then you have to watch out, because somebody, quite possibly you, will have to pay for that good fortune. At one extreme it seemed to me that the death of rock stars is connected to this notion - Hendrix, Joplin, Lennon, Jackson, Houston and the rest. Sure, they did drugs, sure they all lived high on the hog, sure they were half crazy and probably they had delusions, or they were sinners and did bad things, just like the rest of us, but they brought sustained, ecstatic meaning and pleasure to millions of people. Someone had to pay for that, so they paid.

My horse Ginger was my greatest and best friend. He earned my bread and butter. He brought me adventure, respect, excitement, danger, challenge, insight, craft skill, glorious joyous fun, a living, breathing link with two thousand year old culture, and the deep satisfaction of learning to manipulate living natural forces and bring them under my command. He gave me real physical power at my fingertips, the power to make physical change in the real world at the end of a short leather rein, yet he was flesh, blood and bone, and he produced that power from nothing but grass and sunlight. He led me into a wonder world of breathtaking landscapes, hidden wildlife, arcane mysteries, fabulous beauty, ancient traditions, the friendship of some jewels of men and women, and the simple pleasure of his company. The price I had to pay for all that was my grief, and a vague but enduring guilt that I had let him down at the end, simply because his life was in my hands, and I survived him.

With Ginger at Buttermere.

24. The Last Pull - the tipi

With Ginger gone, I had no means of earning a living in the wood. I did not have the heart to buy another horse, but the diversification projects (tutoring at Brathay Hall, Animal Connections, and the embryonic Project Phoenix), were too intermittent to support my family. I decided that I had to get a 'proper' job, and set a deadline of April 1st, 1984. Failing that I would use the last of our meagre savings to try to find a new horse and start again in the wood. My heart was not in it because my experience of Sam Pig, Beauty, and Sally, and even the Fell pony stallion Charlie all convinced me that Ginger was the best timber horse that there ever was, and no other horse could match him. Even if I found another horse, I was weary, wary, and slightly damaged, but I had few options, and if necessary I would have to start again.

I read the local papers every week, searching for any job that would allow me to carry on some of my woodland activities, in particular Project Phoenix, while earning enough to feed the family and pay the rent. Nothing, nothing, and more nothing. Until one bright day, when the job of Music Officer at the Brewery Arts Centre was advertised in *The Westmorland Gazette*.

This job had my name written all over it, in copperplate writing. I put together a new CV, building up my Drama Degree, my business qualification, the Brathay management training, the Arvon Foundation, the Ceilidh band and my regular work as a banjo-playing folk singer in a busy music venue. I got as far as an interview with the dynamic director, Anne Pierson, and the urbane Mark Monument from the Northern Arts Association. The interview went well, I was hopeful, and they turned me down. That was another bad day, and I resigned myself to a return to hard graft and a new young horse. My April 1st deadline was close, so I phoned Geoff Morton to see if he had any suitable Ardennes horses for sale. He did not, but he suggested I talk to the main Ardennes horseman in the UK, Charley Pinney, and I arranged to visit.

The next morning the phone rang. I had expected Charley Pinney but it was Anne Pierson from the Brewery, sweet-talking: 'Hello, how are you?' It turned out that the preferred candidate for the Music Officer job had thought better of it and turned it down. Would I accept the position? (Is the Pope a Catholic?)

Jubilate! I accepted, I celebrated, and I danced a little jig. The page had turned. Everything changed in a day, and suddenly I had to plan for new departures and new horizons. Somehow I would have to manage the transition, and in particular I had to tell Tom and Michael that I could not spend the next 12 months charcoal burning, so I could not run Project Phoenix.

By yet another twist of synchronicity, my dad Walter was emerging from his own domestic crisis at the time, and he was the only free agent I could think of who knew anything at all about charcoal burning, so I called him up, told him the news, and asked if he wanted to run Project Phoenix. He barely paused to think, and just said yes. Within a month he had yoked up his bow-top wagon, with two ponies pulling and three more tied on behind, and shut the door of the farm where he had been for 35 years, for the last time. (He did not lock it, but that is another story.) He headed horse-drawn for the Lake District, and he never went back.

Two weeks later he set up his camp at Thwaite Head in the Rusland Valley, and over the next year or so he produced several tons of charcoal and all the raw data we needed for the Project Phoenix report. By then I was flat-out working as a music promoter at the Brewery Arts Centre, so I kept my distance, but I wrote up the final report, Michael Gee edited it, and we presented it to the National Park officers on time, on budget and right on the money.

We had a first class product, a competitive price, and a list of target customers. Walter had the bit between his teeth, and so Lakeland Charcoal was launched. I taught myself to use project-planning computer software for the first time, so we were not short of plans. We had a business plan, a detailed marketing plan and an operation plan for an efficient ten-day burning cycle for two men using two kilns, working long hours and producing thousands of pounds worth of quality charcoal. I mapped out three sales routes around the Southern Lake District, using standard academic market segmentation. Walter drove each of these routes on three days in succession, calling in at every target retailer (and a few that we had missed) and left each of them a sample bag of charcoal and a leaflet. By the end of the week we had over three thousand pounds worth of orders. We were in business, and making proper money, and I had my new job at the Brewery Arts Centre and a salary to boot. We had proved that reviving charcoal production was not only viable, but represented a solid investment with an excellent return, and I had cash in the bank.

Walter was pleased and happy and found a new lease of life, an Indian Summer at the age of 60, and he lived out in the woods for the next thirty years. After a year or so as a charcoal producer he started running courses in charcoal making, which paid him just as well without the hard graft. About a year later one of his course participants, Simon Grey, bought the Lakeland Charcoal business, and Walter started up another successful venture, Wally's Willows, and slowly grew into his new role as the Old Man of the Woods.

With Walter in the driving seat of Project Phoenix I had no more commitments to my woodlands project, but I had one unfinished project close to my heart, and which allowed me to keep a link to the values which had sustained us for the last seven years. The Tipi project had been brewing for a year or more, and was just becoming a reality.

It was an appropriate, and almost inevitable, coincidence, that the last pull I made with Ginger before he died was the start of the tipi project. Just before his health failed, we felled and extracted 20 tall poles of Lodgepole pine out of Pull Wyke Woods, by the shore of Lake Windermere, to make the poles for a Native American Indian tipi.

I had first encountered a tipi at Megan Fair in Pembrokeshire, when the Tipi People made a dramatic entry with an old GPO Truck. The truck was designed to carry telegraph poles but now carried 20 foot tipi poles, from the well established alternative settlement at Tipi Valley in mid-Wales. The next time I saw one it was being used as the cider tent at Glastonbury Festival, on 7/7/77, where it housed a very large cider barrel. When I poked my head through the tipi door in search of cider there was a man lying on his back under the tap, with this mouth wide open, the amber nectar running straight out of the barrel and down his throat.

Until the Mongolian Yurt became the preferred budget DIY dwelling for those who wanted to settle on the land, the tipi had been the structure of choice. When we first embraced alternative culture at Littlebeck, Tony Ashford had introduced me to the comprehensive and invaluable *Indian Tipi Book,* by Reginald and Gladys Laubin. We had considered building one to live in until, we realised what a serious undertaking the construction would be. Our tatty old caravan was a quicker, easier and cheaper solution to the accommodation problem, but we never lost the ambition to build a tipi, and after seven years, we built one.

We build it with a group of 'alternative society' friends, comprising about 10 like-minded adults and a rake of children. We met at weekends at Sprint Mill, Burneside, drawn together by the same sense of impending economic and environmental disaster which had first taken me into the woods. David Hicks, a professor at the Peace Studies Centre, with his wife Sylvia and their two boys, brought an academic and international world view. The Acland family and their rambling mill full of bygone treasures brought a sense of history and a living tradition, a deep suspicion of the Magnox Nuclear Fuel plant, and a passionate intensity. We were joined by West Cumbrian anti-nuclear campaigners Joanna Treseder and John Uden, and by Hans Ullrich the potter and Mary Barratt (now Ullrich), craftswoman, author and otter-spotter. We came together to share our hopes, dreams and frustrations, and talked constantly of 'The Big One' – a theoretical blockbuster project that would strike a (non-violent) blow for sanity and make a real difference to the mad, mad world.

Our group was politically active, although not on the front-line. We would talk for hours into the night and trained as an NVDA (Non Violent Direct Action) Support Group. We went to Trafalgar Square to rage against nuclear power, and to Greenham Common to help encircle the base. We had weekend camping trips in the Cumbrian hills, and attended fêtes and festivals to make connections and spread our message. We invited guest speakers from the Oxford Research Group to come and teach us about alternatives to war. We baked bread and cakes, sang songs, rowed on the river, played memorable games of rounders through long summer evenings. We supported each other as best we could, and at the heart of our weekend community was the tipi.

At an early meeting we decided that we needed a focus, a practical project, and constructing an American tipi was the unanimous choice. We wasted no time getting started. Copies of the tipi book were ordered, and canvas was bought. The eponymous Lodgepole pine was of course the preferred species for the poles, but they were tricky to find as they needed to be tall and slender, ideally from an un-thinned but overgrown woodland. By the last of the unlikely but un-surprising coincidences which characterised this story, there was a plantation of Lodgepole Pine of the perfect size (30 feet long) growing 100 yards from my barn and stable at Pull Wyke. I convinced the woodland manager that the plantation needed thinning, which it certainly did, and sharpened up the chainsaw.

The Lodgepole pine plantation was overgrown and the trees hung up in the canopy when the felling cut was made, but Ginger hauled them down with ease, and we cut them to length at 24 feet before he pulled them out to the roadside. We borrowed a car transporter trailer, lashed them on so they hung over the back by about 10 feet, tied red warning flags all over them, and hauled them from Pull Wyke to Sprint Mill.

Over the course of about eight months of intermittent weekends, our group peeled the bark off the poles, shaved them down with drawknives, sanded them smooth, and oiled them up with linseed. We located and borrowed an industrial sewing machine and working as a team we laid out the heavy canvas on the kitchen table and sewed the huge cover for an 18 foot diameter tipi, taking great care to get it right.

Eleanor Lloyd aged 3, peeling tipi poles with a drawknife.

We made oak lacing pins to hold it together, and ash pegs to hold it down. It was a mighty cooperative effort, and a wondrous joy to behold when it went up for the first time, in the field by Ginger's stable, right next to the wood where he pulled his last load.

The whole group of 18, adults and children, gathered one weekend to put it up. We built a fire inside, cooked up a communal meal in a huge iron pot, and had a celebratory feast of food we had grown ourselves. The whole group slept in the tipi with room to spare, marvelling at the lightness and delicacy of the space, the intricate patterns of poles and shadows, and the harmony, symmetry and balance which we had created. We watched the stars and planets move past the smoke flaps, and the flickering light of the fire and the moon thrown onto the canvas. We sang sweet songs, drank home made beer and sloe gin, and talked and talked until the dawn chorus brought an ecstatic finale, and we dreamed the sweetest of dreams.

The tipi was a culmination, a distillation, and a living reminder of the ideals that had taken us into the wood. While we were building it, we were very clear that it was a symbolic act as well as a practical one. It represented a right relationship with each other, with the Earth and its resources, and a closeness to the land and its natural processes. It remains for me a powerful memento, a symbol of those great days in the wood – of fierce idealism, sustainable power, grounded effort, true conviviality, inspiring songs and poetry, committed politics, direct action, laughing children and a genuine sharing of the best that life has to offer.

Those tipi poles, which by one last extraordinary coincidence were Ginger's last pull and my last pull on our great adventure, now sit in their place, laid along the ancient beams in Sprint Mill, waiting for another outing. I like to think that they carry a message of hope and endeavour to another generation. Boldness has genius, power, and magic in it. Begin it now.

The last pull: The Tipi at Pull Wyke

25. Looking Back

Following our dreams or our passions did not inevitably lead us to the bright sunlit uplands, but in spite of my mistaken prediction of imminent industrial and economic collapse, I had no regrets. I had learned that if you hold out for your individual version of sanity against a tide of collective madness, you may struggle to realise your dream, and you might not make a fortune, or even make a living. You might never be famous, or own a yacht or a fancy car or a 48-inch TV, but you may not want those things. You may want something else, and find something that you did not even know existed.

Ali and I had very little to show materially for seven years in the wood. We still have a dining table, made from an oak butt which Ginger hauled out of a wood, and we have a share in the group tipi, and the satisfaction of the spin-offs from the Woodmanship Trust and Project Phoenix. We have children and grandchildren, raised with a certain ethos, but otherwise our great adventure made very little difference to the world. I have some scars, I know how to handle heavy horses and how to use them to get wood out of difficult terrain where no machine dare go. I know the mountains and valleys and the people of the Lake District well enough to love most of them. I have a few photographs of the beautiful places we lived, a few good tales to tell and a few good songs to sing. I made some lasting friends, and I helped a few people to follow the same path. If that was all I have to show for seven years' hard labour it is fair to ask: was it worth it?

I make no claim to breaking new ground, or even keeping the 'old ways' alive in the face of mechanisation. George Read of Witherslack and his son Kevin Read deserve that badge. George was snigging with horses before I started and he was still doing it 30 odd years after I packed in. George Newton and the Lenihan family have both built businesses around logging with horses in the same territory, and in this new world of conservation grants and amenity woodlands there are more horses in the woods now than there were when I started in 1978.

Ali and I were fortunate that we started out in good health, with no dependents, no responsibilities, and a network of like-minded friends. We were even more fortunate that when we moved back into the world, back to Babylon, we had no serious injuries, and two fine daughters to look after. I set out to take on the big machines, and looking back it is not surprising that I could not compete in the market place unless I was subsidised by woodland owners. The marketplace still rules, and the machines rule the marketplace. Our planet is in an even less healthy state now than it was then. We are barely waking up as a species from a self-destructive trance of oil-based consumerism, and the evolution of a sustainable global economic model seems further away than ever. It seems with hindsight that the ideal wheel which I wanted so badly to turn was a hopeless illusion, and our attempt to turn it changed nothing, except possibly locally in Cumbria. Yet if I had my life over again I would do the same, and this book is an attempt to explain why.

My mistaken predictions of imminent industrial collapse in 1976 were only one part of the reason. Certainly we were inspired by the spirit of the times and by contact with charismatic individuals, so maybe we were just easily led. I grew up in the sixties, so I was riding a wave of street protest, environmental awareness and revolutionary ideology. Marcuse, Guevara, Dubcek, and the events in Paris in 1968 were the backdrop to our politics, and I wanted to do my part, to add my voice to the chorus of protest and put my shoulder to the wheel. After two years of intense exposure to literature at the Arvon Foundation, I wanted action, not words, and it was literature, as much as ideology, which inspired me. I wanted '*to front only the essential facts of life*', and to follow the example of Cobbett, George Sturt, Ewart Evans, Jack Kerouac and Gary Snyder. At the same time I wanted the pride and satisfaction of making a living by my own hands, by my own craft skill, and the independence of doing it for ourselves, and I loved the excitement of feeling like a pioneer. Sure, I wanted to save the planet, and I wanted to prove a point about low impact living, but I was doing it mostly because I had to do it, because I could not pass it by, because it gave my life some purpose and meaning.

Once I started, my appetite grew. I was hungry for the breathtaking grandeur and beauty of the landscape, the frisson of the dawn chorus and the sound of running water, the smell of hot resin on the breeze, the sound of the wind in the trees and the delicate tracery of the leaves in spring '*that break the light in colours that no-one knows the names of*'. (The Byrds). I became addicted to the adrenalin-rush of horse logging, and to the calm silence of the woods and the peace of still waters. We lived for raising children, raising chickens, digging potatoes and splitting firewood, for wild ecstatic ceilidh dancing, whipping up flying session tunes round late-night camp-fires; for home brewed beer and for my little copper whiskey still, for the peculiar emotional melting when the moon is on the water, for drifting wood smoke and the unmistakable excitement of the open road; for the sharing of it all with each other and our daughters and the friends along the way, and for the fleeting sense of harmony and belonging that I experienced when working as an organic being in an organic world. We felt at perfectly home in the woods and in our caravan and our cottage. Our lives took on a meaning, unattainable in any other way. It was worth it for all that, and that is why I would do it again.

That is all common-place enough, and may seem insufficient to justify our rejection of the straight world of materialism and consumerism, and exile from Babylon. But the natural world of the woodland cannot be judged by those standards. It is simply different from the world of machines, engines, computers, play-stations, telephones, currencies, politics, and shopping. It is different in kind, it is another universe, a subconscious dimension. The horse belongs there, and the horseman belongs with the horse. It may be a delusion to believe that by living in that world we were living in another dimension, but I believed it at the time, because the 'otherness' could be felt, like a warm breeze blowing over my shoulder. I could not measure or record these intuitive perceptions, and they are hard to describe, so I could not 'prove' anything, then or now, but I stick by my naïve responses, subjective and irrational though they may be. Scientific method and rationality has brought us a long way, but so far has shown itself incapable of understanding our consciousness, that most essential characteristic of our species, so it seems obvious to me that there are natural domains inhabited by humans that are well beyond the reach of science.

My belief in the reality of that 'otherworld' has not faded since we moved on from those long days in the woods, although it has become more difficult to achieve that sense of 'otherness'. Mobile phones and the internet dominate our lives, even for pioneers of alternative culture, and there has been a deluge of (sensible) regulation concerning Health and Safety, planning permissions, animal welfare, bio-security, and grazing management. The noise of machines fills the countryside - road traffic, farm-tractors, tree-harvesters, helicopters, aeroplanes, drones, and 4x4 off-roaders. Nevertheless by giving up the TV and computer, binning the mobile phone, and finding a quiet corner of a quiet woodland, it would remain possible 40 years later to dive into the pool of 'otherness' and emerge refreshed.

Our decision to embrace an alternative vision had some profound consequences for the rest of our lives. The things we learned about cottage economy, about life and tradition among the Lakeland farmers, about the rural economy and the principles of broadleaved woodland management, were seeds which germinated and flowered many years later, when we were able to buy our own land and manage it according to our principles. We were fortunate that within a few years of finishing in the wood, government agricultural policy and support turned away from intensive livestock production and towards conservation and amenity, so our convictions led us very early into managing Environmental Stewardship schemes, planting thousands of trees, hundreds of yards of hedgerows, and managing traditional hay meadows, species-rich grassland, wetlands and large areas of common land. The ideals of the Woodmanship Trust reverberated long after our time in the wood, and a whole culture and craft industry sprung up around those ideals. Cumbria Broadleaves, The Wood Education Project, The Coppice Association, The Bill Hogarth Memorial Apprenticeship Trust, The Coppice Co-op, the Woodland Pioneers, Wood Matters and many other craft organisations have since flourished independently. We can look back with real satisfaction that we played our part in a much bigger movement. Our time was not wasted, and it is a great thing still to be welcomed in woodland clearings and round camp-fires all over the Lake District by those who followed after.

Ali and I embraced this alternative dimension with a combination of respect for practical traditions, a preference for cottage economy and a pride in craft skills. Most of all for the close, balanced relationships between us and horses, between us and the woodland, and the garden, and the earth, in which each is at the service of the other. That balance, that harmony, can only be experienced in the real world. This book cannot be a substitute, but it can tell the story as it happened, as an insight for anyone who is interested, and as a signpost for anyone who would like to try it.

The best advice I can give to anyone who wants to know more is just to try it for yourself. If you cultivate the ground, and then plant a seed and nurture it, that seed will grow and bear fruit. This applies to whatever has life – a tree, or a horse or a child, or a partner, or a planet, or an idea – if you love them and respect them and take care of them, they will grow and thrive. If you are careless, or neglect them, or abandon them, or discard them, or abuse them, they will wither and die.

Life can be hard whatever choices you make, and there is no obligation to make it harder, but however hard the work, however bad the pay, however much the straight world may think you crazy, and in spite of accidents, injuries, tragedies, disillusions, breakdowns, failures, disappointment, depression, anxiety, and the pain of partings, you can follow your dream as far as it will take you. All you have to do is start.

There is no guarantee of success, and the idea that 'you can be anything you want to be' is an illusion. You might well regret what you have done, rather than what you have not done, or it could be the other way round, but either way you will have a story to tell.

Before you start, you can read all the books you can find, and ask someone who has travelled that road, but you will only know for sure if you find out for yourself. You may find good luck and inspiration, as we did, and when you find it, you can follow it. You do not know your destination, but you do not need to, because where you are going is not a place, it is a way of being in the world.

It may not be the same for you as it was for us, but with luck, it might take you to worlds you could not even dream about.

Be bold – try it!

Bill Lloyd
The Slough, Docker
7th September 2012
Revised 14th September 2020

26. Bibliography

Books inspired my years in the woods, no doubt about it, partly because I had a university education and I loved my books, but then I had always loved books. Thanks to my mum and dad, I learned to read early, and devoured book after book, from the age of about 8. We had no electricity on the farm until I was 12, so most of my bedtime reading was done by a candle or a Kelly lamp, or under the bedcovers with a torch if one was available. As a teenager my dad Walter put me on to Henry David Thoreau, which set the ball rolling, and University made me hungry for literature, before two years at the Arvon Foundation opened a hundred doors, and the Arvon writers provided signposts for the road I travelled.

It was Alexis Lykiard who introduced me to the Beat Poets, and I became a devotee of Jack Kerouac and Gary Snyder. I carried dog-eared copies of *The Dharma Bums* and Snyder's poetry books in my day bag for months in the wood, and they became heroes. Kerouac wrote a stream of consciousness about Snyder, and Snyder wrote taut and simple verse about the simple life, his time as a monk, and working as a fire lookout in the great American forests. They described how they would pack up their rucksacks with blankets, water, apricots and almonds, and of course their writing materials, and hike up to Desolation Peak. I tried Kerouac's recommended diet (edible and philosophical) and loved it, so 30 years later, when I am out on the road playing music gigs and festivals, I always have my books, and my apricots and almonds. As Kerouac suggests in his description of 'rucksack wanderers', his 'visions of eternal freedom' are a compelling alternative to drudgery and consumerism. As the physical and financial stress of my years in the wood began to take their toll, I thought more frequently of Snyder's cautionary poem *Hay for the Horses.*

Close behind the Beat Poets in my collection was Ken Kesey's epic novel *Sometimes a Great Notion,* the story of a family of 'gypsy loggers', cutting and extracting timber in the vast native forests of Oregon, in conflict with the unions, the bosses, with themselves, and most of all with the unforgiving wilderness. It is a brutal, tender, penetrating account, written in a bold multi-faceted experimental style.

Among my most significant literary discoveries was the poet Basil Bunting. After I got to know his magnificent long poem *Briggflatts,* nothing was quite the same again. He was not just a writer but a man of action – a British Military and Political Intelligence Officer and correspondent for *The Times* in Iran, where he lived among the nomadic mountain tribesmen. Ten years after Eric Mottram introduced me to Bunting's work at Lumb Bank, I took part in the making of an autobiographical film about his masterpiece and twenty years after that I was part of the celebrations for the semi-centenary of its publication. I have read and listened to Briggflatts from start to finish several hundred times, and still discover new dimensions of love, life, poetry, history and music. It has been a thread running through my story for forty years.

Literary preference is highly subjective however, so the following bibliography focuses on more or less technical publications, without which I might never have made a start. The best of them – George Sturt and George Ewart Evans in particular – combine technique and inspiration.

Worthy of special mention is the *Whole Earth Catalog* and its companion volume, the *Whole Earth Epilog.* My father Walter Lloyd had always had a tendency to embrace the 'alternative society', and I recall that when Ali and I lived at Duckworth Farm during the summer of 1974, the

Whole Earth Catalog and recently published *Epilog* were never off the long table. They were shunted up to one end before meals and then dragged out again afterwards. The house was full of hippies, contrarians, musicians, journalists, wanderers, and enthusiasts for strange alternative technologies – people came from all over the country, all over Europe, and all over the world to discuss radical and mildly subversive projects. Some of the visitors were probably outlaws and some of them definitely were. Their eyes and hands would latch on to the disintegrating, well thumbed *Catalogs*, not only for the specific information they contained, but because these books, with their iconic cover photographs of Earth taken from the Moon, focussed our minds and conversations on the extraordinary richness and diversity of human culture in the context of our tragically finite planet. As Steve Jobs remarked, the *Catalog* was the Google of its day, and 40 years later it still sits on my table, falling apart, but still serving a functional purpose and a spiritual *memento mori*. These were the books that I used to get started. Much has been written since, but this is how it was.

General:

Stewart Brand	The Whole Earth Catalog (1971) & The Whole Earth Epilog (1974)
Goldsmith & Allen:	Blueprint for Survival (The Ecologist)
John Seymour	The Fat of the Land and Self-Sufficiency
George Sturt	The Wheelwright's Shop (A Classic. Brilliant.)
William Rollinson	Life and Traditions in the Lake District
Hartley & Ingilby	Life and Traditions in the Yorkshire Dales
Various	A book of Visions: A directory of Alternative Society Projects (London): The Ideas-Pool 1973
Meadows *et al.*	The Limits to Growth 1972 (Club of Rome)
William Cobbett	Cottage Economy (1822)
Satish Kumar	Resurgence Magazine 1973 – present

Horses:

Charles Philip Fox	Circus Baggage Stock – a tribute to the Percheron Horse. This book traces the use of heavy horses in hauling travelling circuses around the USA in the early 20th century. Includes a picture of a 64 horse team. This is the real stuff. Essential reading.
Maurice Telleen	The Draft Horse Primer (USA) At the time (1980) the only comprehensive text available about the care of working draught horses.
George Ewart Evans	The Horse in the Furrow Ask the Fellows who cut the Hay
War Office	Manual of Horsemastership and Animal Transport 1937

Forestry:

Ralph W Andrews	Timber (Logging in the vast forests of British Columbia.)	
BTCV	Woodlands - a practical conservation handbook	
Forestry Commission	Timber Extraction Standard Times	
	Forest Record No 80	Forest Fencing
	Forest Record No 83	Cross Country Vehicles (The competition!)
	Booklet 26	Volume Ready Reckoner
	Booklet 19	Timber Extraction by light tractor
	Booklet 11	Extraction of Conifer thinnings
	Booklet No 8	Aids to working conifer thinnings
Forestry Safety:	Booklets FSC11 to FSC18 (Invaluable chainsaw safety guides)	
C.E. Hart	Guide to Timber Prices and Forestry Costings (Essential)	
Mary Barratt:	Oak Swill Basket-making (Still the best book on the subject)	
S. Douglas et al.	Forest Farming (How to save the planet by silviculture)	
Edlin, H.L.	Woodland Crafts in Britain, 1949	

Music: Top Picks from the soundtrack to my woodland days:

Irish/Celtic:
Paddy Tunney singing *The Green Fields of Canada,* Al O'Donnell singing *The Bunnan Bui,* Kevin Mitchell singing *The Battle of Aughrim,* Seán Jeaic Mac Donncha singing *Una Bhan,* Anything by Planxty, Paul Brady, Dick Gaughan, 5 Hand Reel, Dé Danann, Alain Stivell

Americana: Music from Mud Acres, The Byrds, The Grateful Dead, Emmy Lou Harris,
British: John Martyn, A.L.Lloyd, Anne Briggs, Mike Harding, Ewan McColl

The following books have contributions by Bill and Walter Lloyd:

New Woodmanship Trust:	The Micro-economics of Coppice Management.(Report)
	Lakeland Charcoal (Report)
Brathay Woodlands Project	Apprentice Training Syllabus (it is out there somewhere)
Rupert Sagar-Musgrave:	Appleby Fair - the greatest Gypsy and Traveller gathering
	Introduction by Bill Lloyd
Andrew Connell	"There'll always be Appleby" - Introduction by Bill Lloyd
Walter Lloyd	How to Build a Bow Top Gypsy Living Wagon
	Travels with a Pony
	The Black Art of Charcoal Making (In Preparation)
Alexander Lloyd	Pioneer Jackaroo: (In Preparation)
	The Diaries of a Queensland Horseman 1864
	Edited by Bill Lloyd
Bill Lloyd	The Whispered Waltz - 40 Years as a travelling musician.
	(In Preparation)

Website and Blogs: www.billlloyd.co.uk

27. Appendices

Appendix 1: Yew Tree Tarn photo sequence

Gathering a load of Larch poles. (Note the deep brash)

Extra poles are added to the load by looping the chain round the tips.

Ginger works to voice commands, while the horseman is protected by a tree.

Ginger knows where he is going

Here the load has been snagged on a young oak tree – some poles have gone one side, some the other. The chain must be loosened, taken in front of the tree and re-fixed so that the load can be pulled out. The horse waits while this is done.

Another snag sorted

The next snag is trickier – the momentum of the moving load has taken it the wrong side of a standing tree.

It must be pulled sideways to the right and then 'fish-tailed' round the obstacle

– but first there is a breakage which must be repaired.

Once the breakage is fixed, there is a clear run down to the stack.

Done

Ginger goes back to his field...

... and I go home to Yew Tree Farm Cottage

Charlie at Yew Tree Tarn

Appendix 2: Parts of the harness

Trace Harness.
The simpler snigging gear comprises just collar and hames, backband and trace chains, and a swingletree that drags along the ground, without the back-straps, loin straps and spreader bar.

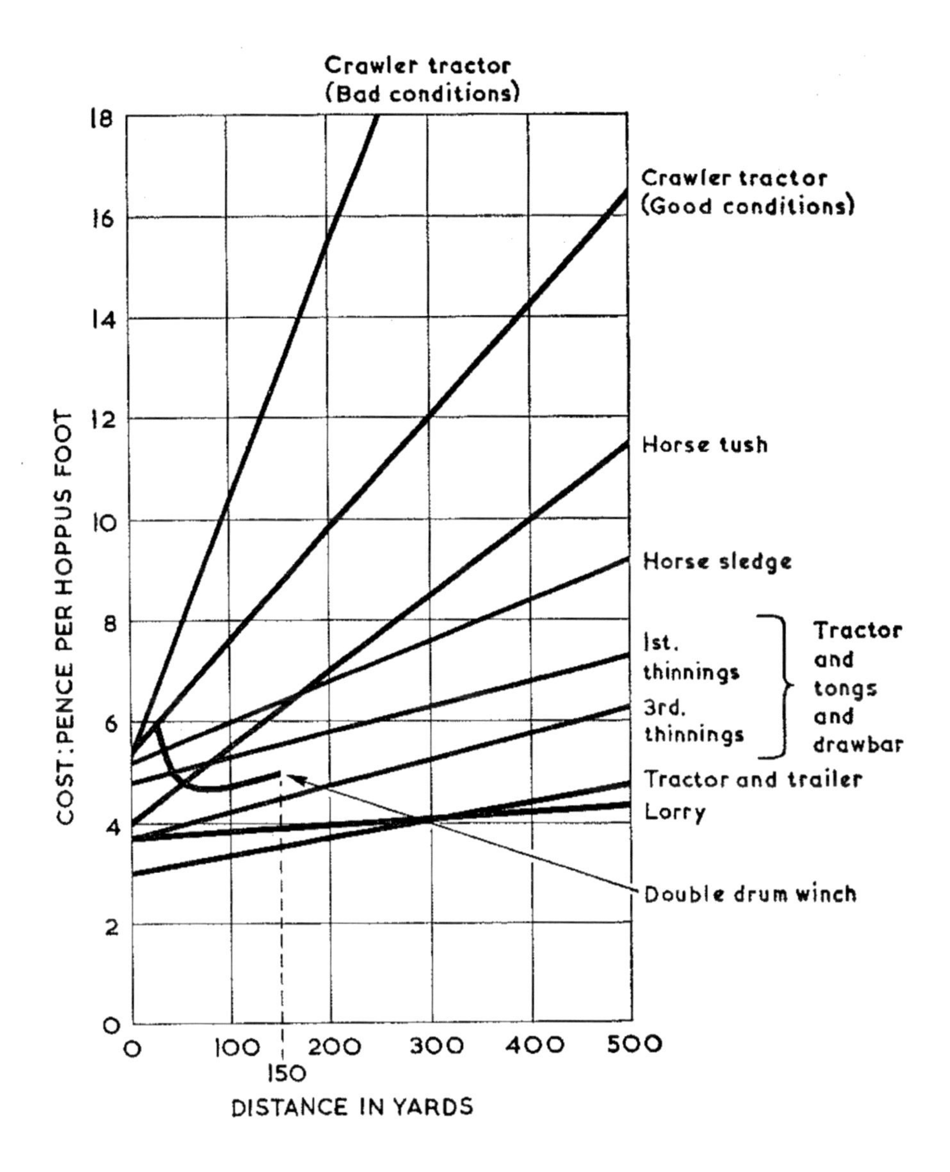

Costs of different extraction methods

Forest Horsepower **December 1983**

Production Record: Chapel House Wood

Example: Earnings from 12 hours horse extraction.
5 Hours Converting and stacking

Item	Number	Weight	Rate	Value	Rate per	Value
	Pieces	Tons	per ton	as tons	piece	as pieces
Posts	329	4.1	£ 7.00	£ 28.70	0.09	£ 29.61
Chip	225	7.5	£ 7.00	£ 52.50	0.24	£ 54.00
6'1" TBM	19	0.64	£ 7.00	£ 4.48	0.24	£ 4.56
5' Pulp	9	0.17	£ 7.00	£ 1.19	0.14	£ 1.26
Totals:	**582**	**12.41**		**£ 86.87**		**£ 89.43**

Summary:		Hours	Rate	Total
			Per Hour	
Horse Extraction		12	£ 4.50	£ 54.00
Woodcutters		7	£ 3.25	£ 22.75
Converting		5	£ 2.50	£ 12.50
Totals:		24	£ 10.25	**£ 89.25**

Average Weekly Wage in 1982	(40 Hour week)	**£**	**154.00**
	Average Hourly rate:	**£**	**3.85**
	Horseman Rate	£	4.50
	Woodcutter Rate	£	3.25
	Convert and Stack Rate	£	2.50

Hourly rates

Appendix 3: Standard Times

Forestry Commission - Horse Extraction of Thinnings

1. **Conditions:** The Standard Times apply to:
   a. Peeled or unpeeled conifer poles, the average volume of which is known and where few if any butt cuts have been made.
   b. The poles should with few exceptions lie in the direction of extraction
   c. Brashing percentage is at least 75%
   d. The floor conditions are average:
      i.. Slopes normally in the direction of extraction
      ii... Fair amount of brash
      iii. Soft peat surface normal
      iv. Frequent turf drains linked by main drains
      v. High stumps caused by turf planting

2. **Work Specification:** The Standard Times are for the following work:
   a. Poles to be extracted with a minimum of damage to the standing trees, precautions being taken as necessary to protect poles by use of brash and tops.
   b. Where drains are filled with brash to facilitate extraction, this should be removed on completion
   c.. Poles to be stacked at roadside, butts together in a tidy manner. Size of bings and direction as determined by the forester.

3. **Tools**
   a. The rates derived from the standard times make allowance for the horse to be owned by the forest worker or to be hired.
   b Hand tongs, Axe, and Bow saw are normally required.

4. **Allowances included in the standard Times:**
   a. Personal needs and rest according to size of tree, of the time spent loading, unloading and travelling in and out: 15 - 20%
   b. Work other than that performed on individual trees or loads and for contingencies; clearing brash, filling and opening drains, attention to horse and harness: 20%
   c. Allowance for extra time to get the horse to the job and the extra half hour a horse needs at lunch time: 12%

5 **Variation in Conditions**
   a. If conditions are more difficult, very soft ground with many drains, steeper slopes or rock, then rates can be increased by up to one eighth. ADD up to 12.5%

6. **Method of Setting a Rate:** The standard times have been found to vary with the size of pole, known from the tariff, and the average length of haul; in addition the rate must vary according to the conditions of service of the man and ownership of the horse.
   a. Average distance of extraction. This is calculated as follows, and presumes that the poles lie evenly over the felling area.
      i. For simple areas: the average tushing distance is the average of all the shortest differences and all the longest distances along which the poles are tushed.
      ii. Thus if a roadside compartment varies in depth from 100 yards to 150 yards, the distance increasing evenly, then the shortest distance in every case is 0 yards, the longest is 125 yards (average of 100 yards and 150 yards)

and the compartment average is 125/2 = 60 yards. As generally the actual distance travelled is longer than the theoretical, payment for a 75 yard haul would be justified.

b. If there is dead ground between the road and the extraction area then the minimum distance is no longer zero. E.g. Nearer edge of the compartment varies from 50 to 100 yards from road. 75 yards average. Further edge of the compartment varies from 150 to 350 yards from road. 250 yards average. Then average for compartment is 75 + 250/2 = 162

c. If the compartment is very variable, it would be better to pace some of the distances actually likely to be used.

d. With experience and knowledge of the area, estimates can be made from the map.

7. Variation in rate for standard minute according to ownership etc.

   a. The rate per standard minute for the horseman as an employee of Forestry Commission will be calculated in Appendix A, (i.e. at February 1960.)

   b. For a contractor the rate should not exceed more than 25% over this, viz. not more than 1.4d (sic.) (i.e. 1.4 old pennies)

Standard times per pole for horse extraction by tushing

Average Pole (Hoppus Foot)	**Average distance in yards**							
	25	**50**	**75**	**100**	**125**	**150**	**175**	**200**
1	1.80	2.00	2.10	2.30	2.50	2.70	2.90	3.10
1.25	2.00	2.20	2.40	2.70	3.00	3.20	3.50	3.70
1.5	2.20	2.50	2.70	3.10	3.40	3.70	4.00	4.30
1.75	2.40	2.70	3.00	3.40	3.80	4.20	4.50	4.90
2	2.60	3.00	3.50	3.80	4.20	4.60	5.00	5.40
2.25	2.80	3.30	3.70	4.20	4.60	5.10	5.50	6.00
2.5	3.00	3.50	4.00	4.50	5.00	5.60	6.00	6.60
2.75	3.20	3.80	4.30	4.90	5.40	6.00	6.50	7.10
3	3.40	4.10	4.60	5.20	5.80	6.40	7.00	7.60
3.25	3.80	4.30	5.00	5.60	6.20	6.90	7.50	8.30
3.5	3.90	4.60	5.30	6.00	6.60	7.30	8.00	8.70
3.75	4.20	4.90	5.60	6.40	7.00	7.70	8.50	9.30
4	4.40	5.20	5.90	6.70	7.40	8.20	9.00	9.80

Standard Minutes

Notes:
a) For distances over 200 yards, add 0.2 S.M. per hoppus foot for every 25 yards
b) Interpolation may be made for pole size if necessary. E.g. average pole 1.9 H Ft., average extraction distance 100 yards, standard time = 3.7

(Forestry Commission NE(E) Section XXII, No 6 1963)

Appendix 4: Top Knots

Anyone who is not an English trucker would probably wonder what on earth a dolly knot was. The word is not in Chamber's concise English dictionary neither is it in the Oxford concise. A dolly knot is a special knot that English truckers use to rope a load onto a trailer. I learned it from Ian Sagar of Bacup, a general dealer, when I rode back with him from Appleby Fair to Whitworth. We had a donkey in the back of his wagon at the time, but that is another story.

The dolly knot works like a pulley, and enables you to get double the tension on the rope compared to a single rope. To double it again you can put another dolly knot in the pulling part, thus multiplying original tension. by four.

Archimedes said "Give me a fulcrum, and a lever long enough, and I will move the world". A Dolly knot can be used the same way.[1]

THE DOLLY KNOT **THE BOWLINE**

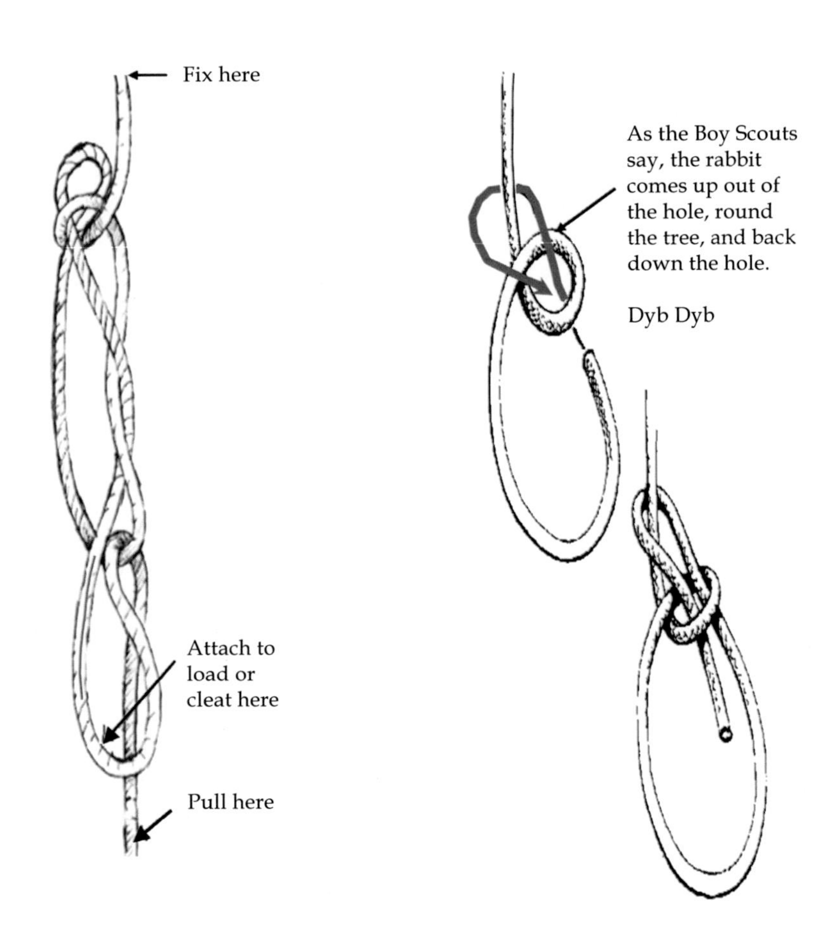

Both these knots need a little bit of practice, but once learned they are surprisingly useful. If you work with horses, hardly a day goes by without using one of them. The main uses of the dolly knot are for fixing a load or tightening rope. The bowline is an extraordinary knot – no matter how hard you pull on the loop, you can always undo the knot afterwards. Good for towing vehicles or catching wild horses. A double bowline can be used as a bridle or halter in an emergency.

28. Glossary

[Note: I have included words which will be well known to woodland or horse workers, but not necessarily to the general reader.]

Ardennes	a compact and powerful breed of draught horse from the heavily forested Ardennes region of France and Belgium.
ATV	All Terrain Vehicle, also knows as a Quad Bike.
axle-box	a steel tube, often tapered, set into the elm nave (or centre) of a wooden wheel which acts as a bearing.
bait	dialect for the workers' daily food, also known as snap, or in East Yorkshire as 'Lowance.
bark barn	oak bark, used for tanning leather, was a primary product of the Cumberland and Westmorland oak coppice woods, It was stored to dry in a bark barn.
Bass	local name for Bassenthwaite.
Beating up	working up the line of young trees, replanting where individual plants have died.
bells	harness bells were worn by some cart-horses to warn oncoming vehicles and people of their approach.
bender tent	a portable shelter made of bent coppice rods covered with canvas.

besom	a broom or brush made of birch twigs bound onto a pole. Now popularised as a 'witches' broomstick.'
BHMAT	Bill Hoggarth Memorial Apprenticeship Trust
bill-hook	a coppicing tool, usually with a double edge, one of which is curved.
bing	this word appears in the Forestry Commission Standard Times. I never did know what it means, and find no reference to it anywhere except as a shale heap.
birch	a common tree species, easily grown, used for bobbins and other turnery. Particularly beautiful and elegant in early spring.
bit	the metal bar which goes into the horse's mouth, held in place by the bridle, and controlled by the reins, which give signals to the horse when pulled. There are many sorts of bit, and I preferred the 'carter's bit' which could be easily removed from the horses mouth without taking the bridle off, by means of a ring and toggle. In fact I often worked without a bit, since Ginger worked to voice command.
bobbin lathe	bobbins were widely produced in Furness, used mostly for the cotton mills of Lancashire
bodhran	an Irish drum.
bowline	the most useful of all knots for the horseman. You should be able to tie this knot easily with your eyes closed, and preferably one handed and behind your back.
Bow-top	horse drawn living accommodation which uses bows bent over and fixed to a dray.
brash and brashing	Part of establishment pnd maintenance of a plantation. Coniferous branches from the trunk, starting at ground level. As the tress grow and canopy closes overhead, these branches die off due to lack of light. 'Brashing' describes the job of cutting off these side branches, up to a height of about 6 feet, using a brashing saw. Brash also refers to all the side branches cut off after a tree has been felled.
Brathay Hall	a Management Training and Adventure Education centre near Ambleside.
breaking	the process of training a horse to work.
broadleaf	type of tree, usually a hardwood, distinct from conifers or softwood.
butt	another word for a sawlog. The thick end, as opposed to the thin end, of a pole.

canopy	the highest part of a tree, a wood or a forest. In a 'closed' canopy, the branches from each tree have grown to meet the branches of its neighbour, so little light can penetrate. An open canopy is necessary for natural regeneration and to allow dominant trees to increase their girth as well as their height.
cant hook	a device for turning and twisting large trees and butts.
cart gear	harness with a saddle and back-chain to support the weight of the shafts. (see also Trace Gear.)
chaff	the husk of wheat, oats or barley, removed from the grain by winnowing.
chillum	a small cannabis pipe
checked area	a part of a wood, often boggy or rocky, where the trees have not grown.
clear fell	complete clearance of a wood, usually done when the trees are mature and before replanting.
click	Westmorland dialect work for a short sharp uphill drag.
Clydesdale	a heavy draught horse from that area of Scotland, bred primarily for carting and ploughing.
cob	a type rather than a breed of horse, the cob is an ideal timber horse. Bigger than a pony (i.e. taller than 14.2 hands) the cob is usually strongly made, with heavy joints, deep chest and a steady temperament. Coloured (i.e. piebald or skewbald) cobs are favoured by Gypsy and Traveller people, who also prefer plenty of feather.
compression wood	when cutting timber under tension, a cut into compression wood will tend to close up and trap the saw. (see also tension wood.)
conifer	softwoods such as spruces, pines, firs and larches.
conversion	the process of making a saleable product out of a felled tree in its natural state. Cross-cutting a pole to a specified product size.
coppice	the process of cutting back a tree to allow multiple shoots to re-grow from the root-stock. The process can typically be repeated at regular intervals of 7-10 years.
a cord	a meaure of stacked wood, Generally taken to be 128 cubic feet of 4 x 4 x 8 feet
cords	long reins, often made of hemp, or woven cotton tape instead of leather.
cross-cut	a saw-cut at 90 degrees to the length of a pole.

Term	Definition
Dales pony	a native British pony breed, a little heavier and larger than the Fell Pony, and almost as hardy, (but not quite.)
dolly knot	also known as a 'Wagoner's hitch', this knot allows a rope to be tightened up by giving a double gearing effect, and can be quickly undone.
drag shoe	an iron or steel slipper which is chained on to a dray or wagon so that it can be easily slipped under a rear wheel and prevent it from revolving when going down a steep hill.
dray	a four wheeled flat bed horse lorry.
dressing out	another word for snedding or taking off the side branches of a tree
driving gear	harness with a leather saddle pad and a leather back band. Attached to 'tugs' through which the shafts are passed. The traces are usually of leather, rather than chain. Lighter than cart harness.
establishment cost	the cost of growing trees. It includes fencing, draining, planting, weeding, 'beating up' and early thinning.
extraction	the process of getting the felled trees from the stump to the roadside.
farrier	a shoeing smith.
feather	the hair which grows around the hoof. A 'full feathered' horse must have thick hair to ground level all around the hoof. This includes the front! Feather is desirable to some breeders, but too much feather can be a real problem for work horses when it gets iced up.
Fell pony	a British native pony breed. Very hardy and strong for its size.
felling	cutting down a tree.
felloe or felly	part of the wooden rim of a wheel.
fire	fire is the major cause of loss in forests. Windblow is the second commonest loss.
fuel ratio	the amount of oil put into the petrol of a chainsaw to lubricate the piston and crankshaft. Although 25:1 and even 50:1 are normal, the more oil you put in the longer your saw will last, but the worse will be the damage to your lungs from breathing smoke.

gang A timber gang would comprise one wood-cutter, one horseman or tractorman, and one man converting and stacking. A trainee or extra might be put on 'roustabout' work, – mending harness, sharpening saws, unblocking drains, or bagging firewood. It would be unusual for all the jobs to go at the same speed, so everyone in the gang should be capable of doing all the jobs.

gaskin the upper part of a horses leg, above the hock.

gear (s) another word for harness.

gelding a castrated male horse. Often preferred for work as they are more reliable (less hormonal) than mares or stallions – but they don't breed.

governess cart also known as a tub-trap, this is a light-weight vehicle for 1-4 people with a door and a step at the back.

half-hitch the most basic knot, in which the rope is tied around itself.

hames metal, or occasionally wooden, fittings which take all the weight of the load and transfer it to the padded collar. The trace chains attach to the hames, which are joined top and bottom by hame straps.

hard feed corn, (oats, maize or barley, or maybe horse-nuts) which is essential for work horses, supplemented by hay or sugar beet.

harrowing dragging discs or chains over grassland or ploughed land in order to break up the surface or spread manure. Also known as 'dragging.'

haylage feed which is made from cut grass which is not quite dry, so is bagged like silage in order to exclude the air and stop it from going mouldy.

HIAB an abbreviation of Hydrauliska Industri AB (Hiab), a Swedish manufacturer of hydraulic cranes.

high pruning similar to brashing, but a longer handled saw is used in order to produce knot-free timber up to 12-14 feet.

hoppus measure an old system of measuring the volume of square sawn timber in a tree, also knows as 'quarter girth.' A special tape is used to measure the circumference (girth) of a butt at the mid point. This is then multiplied by the length, using a pocket ready-reckoner, to give the theoretical volume in Hoppus feet (cubic feet) allowing for taper and bark. See also metric volume.

horse bee the Bot Fly (*Gasterophilus intestinalis)*

horse nuts	compounded horse feed made of milled cereals, mixed with dried grass, oils, possibly carob or soya, minerals and vitamins to make a balanced diet.
horsepower	in order to sell his steam engine, James Watt needed to determine the number of horses his steam engine could replace. This was established by using horses to pull a rope passed over a pulley attached to a weight at the bottom of a deep well. One horse could easily raise a weight of 100 lb (45 kg) while walking at 2.5 miles (4 km) per hr or 220 ft (67.1 m) per min. This accomplished 22,000 ft lb (3041.6 kg m) of work. Watt increased this by 50% to allow for friction in his engine and for good measure, thus establishing 33,000 ft lb (4562.4 kg m) per min or 550 ft lb (76 kg m) per sec as the unit of power or 1 horsepower.
hung up	trees which stay upright even after they have been cut off from the stump, due to a closed canopy.
Irish Draught	a breed of strong work horse, usually 15-16 hands. Originally bred for farm work they are now used as the basic breeding stock for heavy hunters, show jumping and eventing horses.
iron smelting	the coppice woodland of Furness once produced copious amounts of charcoal necessary for iron smelting. The Furness area was once the cradle of the industrial revolution, due in part to its woodlands.
Jackaroo	an Australian word for someone who works on the farm in order to learn the business. They would often be unpaid and have to support themselves, but in return they would get first hand experience as a stock-man in the bush.
jag	dialect for a short steep slope, also used to mean a single load.
knacker	the slaughter man or meat-man who takes old, worn out or fallen livestock.
larch	a coniferous softwood. The only common deciduous conifer.
line thinning	thinning by taking out one row of trees at a time, regardless of the quality of those taken and those left standing.
Lodgepole pine	(*Pinus contorta*) traditionally used in North America for contructing tipis. It will tolerate wet ground, and in UK is used as a nurse for Sitka spruce.
loosing out	unhitching the load, and usually taking off the harness as well, and then tethering the horse or turning it loose.
marra	a mate or a buddy.

MDF	Medium Density Fibreboard. A processed material used in construction and furniture making.
merchant	merchants are the middleman between the woodland owner, the contractor and the end user. They buy standing timber, employ a contractor to fell, extract and convert, and sell the product to a mill.
Merry Neet	a night of songs, stories, pie and pea supper, plenty of beer and good crack. There is often a raffle and an auction of home produce, fancy sticks, and other donations.
metric volume	the modern method of measuring timber volume. The top diameter is multiplied by the length, using a pocket-book ready reckoner. (Probably now superseded by a mobile phone app, but I have not used one).
Norse winch	a tractor-mounted, 3 point linkage forestry winch.
Norway spruce	a common coniferous species, also known as the traditional Christmas Tree. It has red-brown bark and softer needles than the Sitka spruce.
nurse crop	a crop of fast growing trees is planted at the same time as the main crop. The nurse trees will grow faster and higher encouraging the main crop to grow straight in order to reach the light. At first or second thinning the nurse trees are taken out, leaving only the main crop, which has grown tall and straight.
piebald	horse colouring. Piebald is white and black , skewbald is white with any colour other than black. Gypsies call these coloured horses, and in USA they are called pinto horses.
piecework	pay decided by productivity. Usually paid per ton, but occasionally paid per tree, or per fencepost, or per rustic pole etc.
pit-props	timber cut to variable lengths, usually 3 – 6 feet, for use in mining. Now largely replaced by the hydraulic version.
plate, or root-plate	when a tree is blown over by the wind, often on shallow soil, it will fall without breaking away from the spread roots, which will be left standing vertically up in the air.
ploughing team	ploughing was usually done in pairs, but three, four and five horse teams were not unusual.

pole	another word for a stick – a tree which has been felled and dressed out. It could be any size, from a 10 foot rustic pole with a 3″ butt to a 30 foot sawlog. The word was often used as a deliberate understatement so that the 30 foot log was called a 'stick' to indicate that it was no trouble.
pole lathe	a lathe powered by a foot-treadle, and requiring no engine, just a springy pole attached to a string which goes round the work and moves very fast in alternating opposite directions as the treadle is pressed and released.
pulp or pulpwood.	paper is made from wood pulp, produced mostly from softwood trees. The pulp mill uses either mechanical or chemical processes to separate the cellulose and lignin molecules.
ride	a roadway or track way, usually unsurfaced, cut through a stand of timber to facilitate harvesting. Rides may be combined with firebreaks, and can be planned at the time of planting, or cut out during first thinning.
rolled oats	a staple food for horses.
rope	rope comes in various kinds, and rope handling is a useful skill in working with horses. In general braided rope is preferable to twisted rope as it does not kink so easily. Braided climbing rope is usually to be avoided as it is designed to stretch.
rustic poles	a saleable product at garden centres – Straight poles 10 foot long with 3″ butts down to 1″ (approx) tips.
sawlog	timber intended for the sawmill. At least 9″ top, at least 10 feet long, and almost unlimited butt size. Sawlogs are the most valuable part of the timber crop, although in 2012 many sawlogs are going for firewood.
scrow	Westmorland dialect for a mess, a muddle.
Scots pine	a ommon conifer. The wood can be soft to cut when green, and the canopy is less dense than the spruces.
selective thinning	a process whereby the dominant tees are favoured and the dead, smaller and misshapen trees are taken out.

sheaves	when corn was cut by hand or with a binding machine it was a slow process and the harvest might last for weeks, so it was started before the corn was ripe. Once cut, it was gathered into bundles and tied up, first with a twist of straw made on the spot, and later with binder twine. Different types of corn needed slightly different treatment, wheat sheaves being heaviest, then barley and oats lightest.
Shire	a breed of heavy workhorse.
shoeing	essential for logging horses, due to the traction and the danger of hitting bedrock or boulders. A split hoof can take months to heal, during which time you have to feed your horse but he is not feeding you. Shoeing is expensive, but it pays to get the best farrier you can find and stick with him.
silage	grass crop which is either baled or clamped before it has dried. Horses will eat silage, but they usually prefer good hay.
sisal	a natural fibre used for making baler band and rope. Now replaced by polypropylene.
Sitka spruce	(Picea Sitckenis) the most common commercial softwood. It has grey-brown flaky bark and silvery blue green sharp needles. It is resistant to wind and tolerates poor soil. Will grow where nothing else will.
skewbald	a colouring of horses. White with any colour other than black.
skid	skidding is what the tractor men call extracting timber when the load is hauled in to a winch, then raised up and dragged out. A skid is also a short length of timber used as a ramp to build a stack.
skyline winch	a winch common in the 1970s and 1980s, whereby a high wire like a zip-wire is set up from the felling site to the roadside, and a winch is used to hoist the load off the ground and bring it in.
snatch-block	part of a block and tackle, used to gain extra pulling power or to change the direction of pull by passing a winch rope over a pulley.
snedding	another word for dressing out or taking off the side branches.

snigging	a dialect word for timber extraction using horses. See also 'tushing.' Both words are hard to find in dictionaries, although 'snigging' is used in Australia.
snigging chain	the chain used to attach the load to the horse.
snigging gear	the harness used for snigging. It is simpler and different from trace gear, comprising only a collar, back-band and traces. It has certain disadvantages, in that the horse can get tangled in the side chains when stepping sideways, and the swingle-tree can catch on stocks.
specification (spec)	each contract has a spec. which is the product that the merchant has agreed to buy. Trees are usually cut to spec, (converted) at the roadside, although the 'shortwood' system cuts them to spec at the stump. The simpler the spec. and the fewer the products the better for the contractor.
stag	a young unbroken colt.
Standard Times	a Forestry Commission publication which sets out the expected productivity of its work-force. Forestry work is generally not closely supervised, so unless paid for at an hourly rate, standard times are required to set the rate of pay for any given job. Usually inaccurate, and usually to the benefit of the employer!
standing crop	those trees left standing after thinning operations have been carried out. Horse logging does less damage to the standing crop.
stock	another word for the stump left after the tree has been felled. Horsemen like low stocks.
stooks	a system for drying corn in the field by stacking sheaves in a set formation, usually 5 or 7 pairs of sheaves, but varies according to the district and the crop.
strangles	a common respiratory disease of horses, caused by the streptococcus bacterium. It is infectious, and the first sign is usually that the horse stops eating. A nasty discharge from one or both nostrils follows, and the glands on the neck are swollen. Not usually fatal if treated quickly.
stump	another word for the stock, left behind after a tree is felled.

Suffolk	a heavy breed of work horse.
sugar beet pulp	a useful and inexpensive way of getting energy into a work horse. It comes dried in bags, and must be very well soaked before feeding.
swingletree	part of the snigging gears which takes all of the weight of the load, transferring it from the snigging chain to the traces. In snigging gears the swingletree drags along the ground. For workhorse teams, the swingletree does not touch the ground, and is also known as an 'evener' as it takes some of the pressure off the horse collar by moving backwards and forwards as the horse steps, and also evens out some of the strains when two horses are working together. Not to be confused with a spreader bar, which does not take much weight, but is used to keep the trace chains apart, and away from the horse's body.
tail-rake	a rake for collecting all the 'tailings' or odd stems of hay or corn left behind after binding, baling or combining. A tail-rake can be a good way of providing work for a young horse or an inexperienced driver.
TBM	Thames Board Mills – manufacturers of all kinds of compound boards, precursor to MDF.
tension wood	when cutting timber under tension, the cut will tend to open up as the tension is released. (see also compression wood.)
tether	if grazing is in short supply, the horse can be tethered by fixing a chain (never a rope) to a neck-collar with swivel, and then attaching the other end of the chain to an iron pin driven into the ground with a sledgehammer. Old car half-shafts from the scrap yard make good tether-pins. Tethered horses must be checked regularly – twice a day at least.
thinning	the process of taking out some of the trees to open the canopy and allow the standing crop to grow larger.
timber lorry	the lorry that takes the stack to the mill. They usually carry up to 30 tons of wood, and have their own HIAB grab. Most now have a built in weighbridge so that a weigh ticket can be issued on site.
timber tongs	used for handling wood. Hand tongs fit in a sheath on the belt, and are very efficient – they prevent wear and tear on the body and the gloves, and save much bending and gripping. Tractor tongs are used for lifting very large butts.
tip	the thinner end of a pole

trace harness	see diagram. Used for horses working in tandem, and adapted for snigging.
tree harvester	a mechanised tree-combine, which cuts the tree using a hydraulic chainsaw or hydraulic shears, turns it on its side, passes it though a set of blades which cut off the side branches, then measures it and cross-cuts it to specification.
trougher	a good trougher is a horse which eats well. A bad trougher will get tired much earlier in the day.
tushing	Another term for horse extraction used by the Forestry Commission. See 'snigging.'
Woodland Pioneers	An annual week's course in the South Lakes where woodland apprentices are trained in basic coppice techniques.
worming	most horses carry parasitic worms of various kinds, which should be treated under veterinary supervision. Ideally a sample is taken to check what worms are present and appropriate medication is given. Worms can build up resistance to regular treatment, so variation of medication is needed. Warning: The first time a strong dose of wormer is given, the evicted worm population can be quite a shock to behold.
wrinkles	a dialect word for tricks of the trade.
yoking up	harnessing up the horse and attaching a load.

Index

D

E

F

G

H

I

J

K

L

M

N